STRAITS AS SYSTEMS

Geography, Power, and the Future of Global Order

Ali Ayoub

Straits as Systems

Geography, Power, and the Future of Global Order

First published by Ayoub Sciences Press 2026

Ayoub Sciences Press is an independent scholarly imprint dedicated to publishing rigorous, interdisciplinary works at the intersection of science, strategy, and global systems. The press focuses on high-quality nonfiction, academic research, and forward-thinking analysis designed to support resilient, future-ready institutions.

First edition

ISBN: 978-1-972033-24-1

For those who build systems, question assumptions, and refuse to accept the world as static. Your curiosity is the force that widens every narrow place.

"Where land narrows, civilizations meet; where waters tighten, power declares itself."

Ali Ayoub, Ph.D.

Contents

Preface

The world is held together by narrow places. We tend to imagine global order as something vast—oceans, continents, markets, alliances—but its continuity depends on corridors measured in miles, not thousands of them. A strait is a reminder that the global system is not an abstraction. It is a physical architecture, a set of constraints, and a series of vulnerabilities through which the metabolism of civilization must pass.

This book began with a simple observation: every major geopolitical crisis of the last decade eventually converged on a chokepoint. Energy shocks, supply-chain failures, cyber disruptions, great-power signaling, climate-driven migration—each one found expression in a narrow maritime corridor. Yet the frameworks we use to understand these events remained fragmented. Analysts spoke of shipping lanes, naval posture, energy security, or international law, but rarely of the system that binds them together.

Straits as Systems is an attempt to build that missing architecture.

It argues that straits are not geographic curiosities or isolated vulnerabilities. They are structural components of the global order—interfaces where geophysics, military geometry, economic flows, governance regimes, civilizational identities, and technological infrastructures converge. To understand a strait is to understand how the world organizes itself, how it fails, and how it adapts.

This book was written during a period when the fragility of narrow places became impossible to ignore. The Red Sea crisis reshaped global shipping patterns. The Strait of Hormuz once again demonstrated how a single corridor can influence the price of energy, the stability of governments, and the tempo of great-power competition. The Taiwan Strait compressed military, technological, and civilizational stakes into a single contested channel. The Arctic began to open. The Suez Canal reminded the world that engineered geography can fail as easily as natural geography.

These events were not anomalies. They were signals.

The **Straits Systems Architecture** (SSA) model presented in this book is my attempt to interpret those signals. It is a framework for understanding how narrow places function as systems—dynamic, interdependent, and deeply consequential. It is also a framework for thinking about resilience: how states, industries, and societies can navigate a world where the most important pressures accumulate in the tightest spaces.

This book is not a prediction of crisis, nor is it an argument for fatalism. It is an invitation to see the world differently—to recognize that the stability of the global system depends on understanding the architecture of its narrowest points. If we can see straits clearly, we can see the world clearly.

And if we can see the world clearly, we can shape it with intention rather than react to it in surprise.

— Ali Ayoub, Ph.D.

Decatur City, State of Illinois, U.S.A

07 April 2026

Neutrality Statement The analyses and frameworks in this book are intended to illuminate structural features of global chokepoints and systemic risk. They do not represent political positions, policy preferences, or critiques of any government or international actor. Geographic and institutional references are used exclusively to illustrate system behavior. No inference should be drawn regarding the author's views on the legitimacy, intentions, or conduct of any state.

Introduction

The Return of the Narrow Place

In the first months of 2026, the global economy experienced a shock that should have been impossible. A narrow maritime corridor in the Persian Gulf—roughly thirty-three kilometers wide at its broadest and far narrower in its navigable channels—effectively closed to large-scale commercial traffic. Oil production across several Gulf states fell by more than nine million barrels per day. Within weeks, spot freight rates on Asia–Europe routes had doubled, insurance markets for tankers had seized, and governments from Seoul to Berlin were scrambling to secure alternative supplies. The event was not a war, nor a natural disaster in the conventional sense. It was the demonstration, once again, that the architecture of global order remains anchored in narrow places.

This book begins from a simple but increasingly urgent observation: the world has not outgrown its chokepoints. It has only made them more consequential. For much of the late twentieth and early twenty-first centuries, a powerful intellectual consensus held that globalization, containerization, and digital connectivity were rendering geography less relevant. Distance would shrink, borders would blur, and the friction of physical movement would be engineered away. The dominant image was of a flat, borderless, and ultimately decentralized world in which capital, data, and goods could circulate with minimal regard for the constraints of terrain or sovereignty.

That image was already under strain before the kinetic shocks of 2025 and 2026. The COVID-19 pandemic served as the first systemic stress test of the hyper-optimized, just-in-time global supply chains that had been constructed on the assumption of frictionless movement. When factories closed, ports seized, and vessels were denied entry, the underlying fragility of a system organized around extreme concentration became impossible to ignore. The pandemic did not create new chokepoints; it revealed how thoroughly the existing ones had been underestimated. The military and geopolitical disruptions that followed—most dramatically the effective closure of the Strait of Hormuz in early 2026—did not arrive as bolts from the blue. They arrived as the second act in a drama whose first act had already exposed the limits of the flat-world thesis.

The mathematics of complex networks had long indicated what political rhetoric obscured: global trade does not form a decentralized web. It forms a scale-free network in which a vast number of connections depend on a tiny number of highly concentrated hubs. The more the world optimized for efficiency, the more it concentrated risk. The chokepoints of the twenty-first century are not relics of an earlier age of sail and empire. They are the inevitable products of optimization itself.

Some of these points of concentration are inherited from geology—the narrow, current-swept passages of Hormuz, Malacca, or the Bosporus. Others are human creations, engineered corridors whose vulnerability to climate stress has proven at least as consequential as any natural constraint. The Suez Canal and the Panama Canal, both products of nineteenth- and early-twentieth-century ambition, have repeatedly demonstrated that constructed chokepoints can generate systemic risk on a scale comparable to their geological counterparts. The distinction between inherited and engineered corridors is therefore not merely historical; it is central to understanding how states and societies are now forced to adapt to vulnerabilities they themselves helped to create.

The events of 2025 and 2026 made this structural reality impossible to ignore. The effective closure of the Strait of Hormuz, the sustained disruption of the Red Sea and Suez corridor that began in late 2023, record levels of military activity around the Taiwan Strait, and the accelerating militarization and commercial opening of Arctic routes through the Bering Strait all pointed to the same conclusion: narrow maritime places remain the decisive points at which global systems can be stressed, leveraged, or broken. They are not marginal features of the international order. They are among its central operating mechanisms.

This book offers a systematic account of why that is so. It proposes that chokepoints must be understood not as isolated geographic curiosities or tactical military objectives, while also recognizing that some of the most consequential among them are human artifacts whose risks are amplified by climate change and technological hybridization. It develops a framework—**the Straits Systems Architecture, or SSA model**—that treats each strait as a dynamic system composed of six interacting dimensions: geophysical constraints, strategic-military geometry, economic and bio-industrial flows, governance regimes, demographic and civilizational interfaces, and technological infrastructures. This model shows how changes in any one dimension ripple through the others, producing effects that cannot be predicted from geography or military capability alone.

The SSA model is then used to develop a broader theoretical

claim: that chokepoints are best understood through what I call the Unified Theory of Chokepoints. This theory rests on six interlocking principles—structural compression, dependency asymmetry, systemic interdependence, multi-domain hybridity, temporal dynamism, and civilizational encoding. Together these principles explain why narrow places have persisted as sites of power and vulnerability across radically different technological and political eras, and why they are likely to become even more significant in the decades ahead. The theory also makes visible a structural tension that runs through the entire system: the physical corridors and their ecological and sovereign risks are overwhelmingly located in the Global South, while the financial, insurance, and regulatory architectures that govern movement and absorb or displace those risks remain heavily concentrated in the Global North. This asymmetry is not incidental. It is constitutive of how the contemporary chokepoint system distributes costs and benefits.

The argument is not only analytical. It is also philosophical. Narrow places expose a fundamental tension in modern conceptions of order. Much of contemporary political and economic thought continues to treat openness, speed, and frictionlessness as natural or desirable states, and constraint as an obstacle to be overcome. Chokepoints invert this hierarchy. They demonstrate that constraint is not the opposite of freedom or efficiency but their precondition. They reveal that global systems achieve coordination not through the elimination of friction but through its concentration at specific points. The digital desire for instantaneity collides, repeatedly and inescapably, with the geophysical reality of steel hulls, ocean currents, and narrow channels. This ontological clash between digital speed and physical drag is not a temporary inconvenience.

It is a permanent condition of globalization—one that is now pushing statecraft both upward into orbital space, where Low Earth Orbit constellations are becoming both resilience tools and congested chokepoints in their own right, and inward into the digital control systems of ports and vessels, where operational technology vulnerabilities create new vectors of hybrid risk.

The book therefore advances a third claim, one that is at once strategic and ethical. The global chokepoint system operates as a kind of commons—vital to the functioning of international trade and communication—yet the risks it generates are systematically displaced onto particular places and populations. When a shadow fleet tanker spills oil in a narrow strait, when a cyber attack severs undersea cables, or when climate-driven disruption forces rerouting that devastates local fisheries, the costs are borne disproportionately by the states and communities that host these corridors. The benefits of frictionless global movement are socialized; the consequences of its failure are localized. This spatial Tragedy of the Chokepoint Commons is not an accident of geography. It is a structural feature of a system in which physical risk and financial power are distributed according to different logics—one rooted in the location of straits, the other in the location of capital, insurance markets, and institutional authority.

The book is organized in three parts. The first part develops the conceptual foundations of the SSA model and the Unified Theory of Chokepoints, situating both within longer histories of maritime thought and contemporary debates about networks, resilience, and hybrid warfare. The second part applies the framework to the world's most consequential straits, examining each as a distinct configuration of the six subsystems while tracing their interconnections within the larger global system.

The third part moves from analysis to synthesis, exploring how states are adapting their instruments of power to a chokepoint-centric world, how resilience can be operationalized across multiple domains, and what philosophical and political implications follow from recognizing narrow places as the decisive sites of global order.

The approach is deliberately multi-disciplinary and multi-scalar. It draws on strategic studies, international political economy, network science, environmental history, and political philosophy without being reducible to any single one of them. It treats chokepoints as simultaneously material and symbolic, technical and political, local and systemic. It acknowledges that the data available for such an inquiry—AIS tracking, insurance pricing, military activity reports, commodity flows—are uneven and often incomplete, and that any claim to comprehensiveness must be tempered by an awareness of what remains obscured or contested. It also recognizes that the model itself has limits: it is better at revealing structural patterns than at predicting precise outcomes in highly contingent political environments, and it privileges systemic analysis over the granular reconstruction of any single crisis.

The stakes of this inquiry are not academic. The chokepoint century will be defined by the interaction of accelerating climate change, intensifying great-power competition, rapid technological hybridization, and the restructuring of global supply chains. In such a world, the capacity to understand, anticipate, and shape the behavior of narrow places will be a central determinant of strategic advantage, economic resilience, and political legitimacy. States that treat chokepoints as technical problems to be managed will find themselves repeatedly surprised by their political and symbolic force. States that treat them as purely symbolic or historical phenomena will find themselves outmaneuvered by actors who understand their material and systemic logic. The narrow places of the world are not simply sites of risk. They are sites where the future of global order is being decided.

A final horizon point should be stated explicitly. The Unified Theory of Chokepoints is necessary because the global economy remains bound—materially and politically—to physical trade and concentrated corridors. A true black-swan transition toward localized molecular manufacturing, advanced synthetic biology, or other forms of post-maritime production could one day reduce oceanic throughput and diminish the systemic centrality of straits. That possibility does not weaken the framework; it locates it historically: a theory designed for the era in which global order is still built on ships, corridors, and constrained geography.

This book is an attempt to provide the conceptual tools necessary for that decision. It is written in the conviction that the most consequential features of the international system are often the least visible in conventional maps of power—precisely because they are narrow, compressed, and therefore easily overlooked until the moment they are not. The task is not to eliminate these points of concentration, which is impossible, but to understand them well enough to navigate the constraints they impose and the possibilities they create. In a world organized around narrow places, that understanding is not a luxury of theory. It is a condition of responsible statecraft.

1

Chapter 1

What Is a Strait? A Geopolitical Definition

A strait is a narrow maritime corridor connecting two larger bodies of water and separating two landmasses. The definition appears simple, but its strategic implications are profound. A strait is not merely a geographic feature; it is a structural constraint that shapes how states, markets, and civilizations move through the world. It is the point where the continuity of the ocean is interrupted, where the fluidity of maritime space is forced into compression, and where the global system becomes legible through its narrowest channels.

The geopolitical significance of a strait emerges from this compression. When maritime space narrows, power concentrates. The geometry of a strait determines who can pass, who can block, who can observe, and who can project force. It shapes the cost of trade, the vulnerability of supply chains, and the escalation dynamics of military competition. A strait is therefore not simply a passage; it is a mechanism of ordering. It structures the incentives of states, the behavior of markets, and the strategic imagination of entire regions.

Traditional definitions emphasize physical characteristics—width, depth, currents, navigability. These matter, but they do not capture the full strategic reality, which is measured by dependence. In the first half of 2025, the Strait of Hormuz carried an average of 20.9 million barrels per day of oil and petroleum products—roughly one-fifth of global petroleum liquids consumption and about one-quarter of seaborne traded oil. The Strait of Malacca, among the world's busiest commercial waterways, recorded more than 102,500 vessel transits in 2025 and channeled a substantial share of Indo-Pacific energy and containerized trade. Even an engineered corridor like the Suez Canal demonstrates the logic of compression: when disrupted, as in the 2021 *Ever Given* grounding and the subsequent era of Red Sea insecurity, rerouting adds weeks to supply chains, raises freight costs, and converts a localized event into systemic price pressure.

By 2026, the world is experiencing simultaneous stress across multiple narrow places—Hormuz tension, persistent Red Sea disruptions, heightened military activity in the Taiwan Strait, and accelerating Arctic transit through the Bering Strait. These events are not isolated. They reveal the structural centrality of straits in an era defined by climate stress, great-power competition, and supply-chain fragility.

A strait is also a legal space governed by international conventions, a political space shaped by territorial claims, an economic space through which critical flows must pass, and a cultural space where civilizations have historically encountered one another. It is a multi-layered interface where natural and human systems converge—migration routes that have turned certain straits into sites of humanitarian crisis and civilizational encounter, and digital infrastructures where undersea cables, satellite networks, and cyber vulnerabilities now overlay physical geography.

The Compression of Maritime Space

Oceans are expansive and permissive; they diffuse power. Straits are narrow and restrictive; they concentrate it. Oceans allow dispersion; straits enforce proximity. Oceans enable avoidance; straits compel interaction. In this sense, straits are the antithesis of open water. They are the places where maritime freedom becomes conditional, where geography asserts itself most forcefully, and where the global system reveals its points of leverage and fragility.

A geopolitical definition of a strait must therefore move beyond physical description. It must account for how physical geometry concentrates power and systemic risk—and how this concentration shapes state behavior.

A strait becomes a chokepoint when its closure, or even the credible threat of closure, can alter the distribution of power by disrupting critical flows. It functions as a corridor when its openness enables the movement of goods, energy, people, and ideas at scale. It operates as a frontier when it marks the boundary between political orders or civilizational spheres. It serves as a fulcrum when control over it can shift regional or global balances.

These roles often overlap. A single narrow place can operate as chokepoint, corridor, frontier, and fulcrum simultaneously, depending on conditions of crisis, governance, and technological change.

The Strait of Hormuz is a chokepoint for global energy flows, a corridor for Gulf economies, a frontier between rival security architectures, and a fulcrum of regional power. The Taiwan Strait is a frontier between political systems, a corridor embedded in East Asian production networks, and a fulcrum of great-power competition. The Bosporus is a corridor for Black Sea commerce, a frontier between continents, and a fulcrum of historical empires. The Danish Straits show how a stabilized governance framework can preserve corridor functions for trade and Europe's evolving energy system while still operating as a strategic valve under NATO's security architecture.

This multiplicity is what gives straits their strategic density.

Narrow Places Across Time

The strategic logic of narrow places is not a modern invention. Long before the contemporary era, early modern maritime trade networks transformed straits into critical chokepoints and port cities into the first workshops of globalization.

In the Indian Ocean, the Melaka polity built a thalassocracy centered on control of the Strait of Malacca, while Portuguese, Dutch, and regional powers contested the same corridor for the spice trade. Hormuz functioned as a contested fulcrum between Portuguese, Ottoman, and Persian interests, linking the Persian Gulf to wider silver, spice, and slave networks.

Maritime empires—from the Portuguese and Dutch VOC to the Ottoman, Ming, and Qing worlds—were built on the command of narrow maritime corridors rather than on vast land territories alone. These networks were polycentric and cross-cultural, driven by Asian, Arab, African, and European merchants alike, and they produced early global exchanges of goods, pathogens, ideas, and power.

The same patterns of compression, leverage, and civilizational encounter that define today's straits were already visible centuries ago; what has changed is scale, technological intensity, and the depth of global interdependence.

The Straits Systems Architecture

A comprehensive understanding of straits requires a framework capable of capturing this density. The Straits Systems Architecture (SSA) model introduced in this book provides such a framework. It examines each strait as a dynamic system composed of six interacting subsystems:

- **Geophysical determinants** shaping navigability and environmental constraints.
- **Strategic-military geometry** determining access, surveillance, and deterrence.
- **Economic and bio-industrial flows** carrying energy, trade, and resources.
- **Governance regimes** regulating passage and resolving disputes.
- **Demographic and civilizational interfaces** embedding identity, migration, and cultural meaning.
- **Emerging technological infrastructures** overlaying digital, cyber, and autonomous systems onto physical corridors.

This systems approach treats straits not as static features but as evolving structures whose significance shifts as global conditions change.

This chapter establishes the conceptual foundation for the analysis that follows. It defines straits not as isolated geographic curiosities but as structural components of the global order—narrow places through which the world's metabolism flows and where its vulnerabilities surface. The chapters that follow build on this foundation, applying the SSA model to the world's most consequential straits and showing how these corridors shape the strategic landscape of the twenty-first century.

2

Chapter 2

The Architecture of Chokepoint Power

Power accumulates where movement is constrained. This is the foundational principle of chokepoint dynamics. A chokepoint is not powerful because it is narrow; it is powerful because the world depends on what passes through it. The strategic value of a strait emerges from the asymmetry between its physical scale and the magnitude of the flows it governs. A corridor only a few kilometers wide can influence the stability of energy markets, the viability of supply chains, and the escalation thresholds of regional powers. In this sense, chokepoint power is a structural phenomenon: it arises from the geometry of the global system itself.

In an era of profound geopolitical uncertainty, states require strategic anchors to continuously assess their position, advantages, and vulnerabilities. The study of chokepoints provides exactly such an anchor. It offers a practical framework for understanding how narrow corridors can be leveraged or contested, enabling policymakers to craft realistic action plans grounded in the actual configuration of global flows rather than abstract assumptions. The architecture of chokepoint power therefore serves not only as an analytical tool but as a geostrategic compass for navigating an increasingly fragmented and contested international order.

The Three Mechanisms of Chokepoint Power

The architecture of chokepoint power rests on three interlocking mechanisms: compression, dependency, and optionality.

- Compression refers to the physical narrowing of maritime space, which forces ships, data cables, and naval forces into predictable trajectories.
- Dependency refers to the degree to which global systems rely on these trajectories for their functioning.
- Optionality refers to the availability—or absence—of viable alternative routes.

When compression is high, dependency is high, and optionality is low, a strait becomes a strategic fulcrum. Control over it, or even the credible threat of closure, can shift the balance of power.

Compression is the most visible mechanism. It is the physical fact that makes chokepoints legible on a map. In the Strait of Hormuz, the navigable channel narrows into two 2-mile-wide traffic lanes separated by a buffer zone, with inbound and outbound traffic forced into predictable arcs in close proximity to littoral surveillance and coastal fires. In the Strait of Malacca, the passage tightens to less than 2 kilometers at its narrowest point near Singapore, channeling over 102,500 vessel transits in 2025—an 8.7 percent increase from 2024—into a single, high-density corridor. These constraints do not merely limit space; they create focal points for observation, interdiction, and escalation. In a compressed environment, deviation is difficult and detection is immediate. A single incident can therefore trigger disproportionate effects.

In the Taiwan Strait, the narrow geography places commercial shipping lanes and military patrol routes in constant proximity, turning routine transit into a perpetual theater of observation and potential confrontation. Compression here is not just a cartographic feature; it is a permanent condition of strategic intimacy.

Compression also carries ecological externalities that are often overlooked. Because narrow corridors concentrate maritime traffic, they also concentrate environmental damage: higher risks of oil spills, concentrated emissions from idling vessels, and intensified underwater noise that disrupts marine ecosystems. The very geometry that creates strategic leverage simultaneously creates ecological fragility.

The Panama Canal offers a stark illustration of how climate change can weaponize environmental constraints against even engineered chokepoints. Between 2023 and 2025, severe drought lowered water levels in Gatun Lake, forcing the Panama Canal Authority to impose strict transit restrictions and reduce daily vessel passages. What was once a reliable engineered corridor became a climate-constrained bottleneck, exposing how environmental stress can paralyze critical infrastructure and force global supply chains into costly, longer alternatives.

Dependency magnifies these effects. The more a system relies on a particular strait, the greater the leverage embedded in that strait. In the first half of 2025, the Strait of Malacca carried approximately 23.2 million barrels per day of oil—about 29 percent of global maritime oil flows and roughly 22 percent of total world oil demand—while the Strait of Hormuz moved 20.9 million barrels per day, accounting for approximately 20 percent of global oil consumption and about one-quarter of all seaborne traded oil. These flows are not marginal; they underpin the energy metabolism of entire regions, with East Asia absorbing the majority of both corridors' throughput. Disruption here is not a local inconvenience but a structural shock to markets, supply chains, and industrial output.

The same logic extends beyond energy. The Taiwan Strait sits astride technology and semiconductor supply chains: Taiwan produces more than 90 percent of the world's most advanced chips (below 10 nanometers) and roughly 60 percent of global foundry capacity. Dependency transforms straits into strategic liabilities and strategic assets simultaneously: the higher the dependence, the lower the tolerance for disruption, and the greater the coercive value of credible interference.

Optionality determines the degree to which dependency can be mitigated. When alternative routes exist—whether through other straits, overland pipelines, or diversified supply chains—the leverage of a chokepoint diminishes. When alternatives are limited or nonexistent, the chokepoint becomes a structural constraint.

The Suez Canal and Bab el-Mandeb corridor offer a clear demonstration. During the Red Sea security crisis of 2023–2026, Houthi attacks—enabled by Iranian support—forced major carriers to reroute around the Cape of Good Hope, often adding 10–14 days to Asia–Europe transits, materially increasing fuel consumption, and driving freight rates sharply higher. Insurance premiums for Red Sea transits rose significantly during 2024, and sustained insecurity through 2025 continued to impose risk pricing and delay costs across global supply chains. Yet the Cape route, though costly, remained viable.

In contrast, the Strait of Hormuz has no comparable maritime bypass for Gulf oil exports. Saudi Arabia and the UAE have developed limited backup pipelines (the East–West pipeline to Yanbu and the Habshan–Fujairah pipeline) with a combined capacity of approximately 2.6 million barrels per day—only a fraction of Hormuz's normal throughput and insufficient for many critical non-pipelineable cargos. The Taiwan Strait similarly lacks substitutes for the just-in-time semiconductor flows that sustain global technology supply chains.

Over the longer term, the global energy transition is beginning to reshape optionality itself. As countries accelerate the shift toward renewables, electrification, and diversified supply chains, dependence on traditional fossil-fuel corridors such as the Strait of Hormuz is expected to decline, gradually reducing the strategic leverage these chokepoints have historically provided to both exporters and transit states. Optionality is therefore a measure of strategic resilience: it determines whether disruption is an inconvenience or a systemic shock.

Interdependence and Systemic Risk

These three mechanisms—compression, dependency, and optionality—do not operate in isolation. They reinforce one another. High compression amplifies dependency by making flows more visible and harder to reroute. Low optionality locks in that dependency, turning geography into leverage. The architecture of chokepoint power can therefore be read as a simple structural condition: concentrated trajectories plus concentrated dependence plus limited alternatives.

This configuration manifests differently across straits. In Hormuz and Malacca, high compression combines with extreme dependency and minimal optionality to create near-total structural constraint. In the Bab el-Mandeb and Suez corridor, higher optionality through the Cape route makes disruption costly but containable. In the Taiwan Strait, compression and dependency are acute in the semiconductor domain, yet political and military optionality remains contested. In each case, the interaction of the three mechanisms determines whether a strait functions as a manageable corridor or a systemic vulnerability.

Bypass geographies reveal the hidden costs of low optionality. When primary arteries are disrupted, secondary routes absorb massive systemic strain. The Cape of Good Hope, for instance, saw a surge in traffic during the Red Sea crisis, driving up fuel consumption, crew fatigue, and insurance costs across the global fleet. Similarly, the Northern Sea Route and the proposed India–Middle East–Europe Corridor are being positioned as long-term alternatives, yet both carry logistical, environmental, and geopolitical limitations that prevent them from fully replacing traditional chokepoints in the near term. These bypasses demonstrate that optionality is rarely cost-free; it redistributes pressure rather than eliminating it.

Understanding this dynamism requires a systems perspective. A strait is not a single point of vulnerability; it is a node in a network of interdependent systems. Its power derives from the interaction of physical geography with political decisions, economic flows, and technological infrastructures. The Straits Systems Architecture (SSA) model introduced in this book provides a method for analyzing these interactions. It shows how chokepoint power is produced, how it is contested, and how it can be disrupted or reinforced—revealing how compression operates primarily through geophysical and technological subsystems, how dependency is shaped by economic and bio-industrial flows, and how optionality is governed by strategic-military geometry and governance regimes.

Continuous multinational naval presence at Hormuz—through the U.S. Fifth Fleet, U.S. Coast Guard cutters, and European initiatives such as EMASoH and Operation AGENOR—illustrates how states attempt to manage optionality through governance and military subsystems. Regional variations in piracy and insecurity—from opportunistic theft in the Malacca Strait to organized hijackings in the Gulf of Guinea and the relative stability off Somalia—demonstrate how different configurations of the three mechanisms produce uneven security outcomes.

This chapter establishes the conceptual foundation for the analysis of individual straits. It explains why narrow maritime corridors exert disproportionate influence over global systems and why their significance cannot be understood through geography alone. The chapters that follow apply this architecture to the world's most consequential straits, showing how each embodies a distinctive configuration of compression, dependency, and optionality—and how these configurations shape the strategic landscape of the twenty-first century. As climate policy, technological innovation, and multipolar realignments continue to unfold, the relative importance of individual chokepoints will evolve: some will lose salience as dependence declines, while new corridors and vulnerabilities emerge.

3

Chapter 3

The Straits Systems Architecture (SSA) Model

Understanding straits as strategic organisms requires a framework capable of capturing the full spectrum of forces that converge within them. Geography alone cannot explain their behavior. Neither can military doctrine, economic analysis, or legal interpretation in isolation. Straits are multi-layered systems in which natural constraints, political authority, economic flows, demographic pressures, and technological infrastructures interact continuously.

The Straits Systems Architecture (SSA) model introduced in this chapter provides a method for analyzing these interactions. It is a structural framework designed to reveal how chokepoint power is produced, how it evolves, and how it shapes the global order.

The Six Subsystems of the SSA Model

The SSA model is built around six interdependent subsystems: geophysical determinants, strategic-military geometry, economic and bio-industrial flows, governance regimes, demographic and civilizational interfaces, and technological infrastructures. Each subsystem captures a different dimension of a strait's strategic significance. None is sufficient on its own; together, they form a comprehensive architecture for understanding the dynamics of narrow maritime spaces.

Geophysical Determinants. Every strait begins with the physical world. Its width, depth, bathymetry, currents, tidal patterns, and seasonal variability define the baseline conditions under which all other systems operate. These attributes determine navigability, influence shipping density, shape naval maneuverability, and constrain the deployment of maritime infrastructure.

They also evolve over time. Sedimentation, erosion, tectonic activity, and climate-driven sea-level changes can alter the geometry of a strait, modifying its strategic value. In the Bering Strait, accelerating Arctic sea-ice loss extended the navigable season from a few summer weeks to several months by 2025, transforming a once-marginal corridor into a viable gateway for emerging trans-polar routes.

Geophysical compression also carries ecological externalities that are often overlooked. Because narrow corridors concentrate maritime traffic, they also concentrate environmental damage: higher risks of oil spills, concentrated emissions from idling vessels, and intensified underwater noise that disrupts marine ecosystems. The very geometry that creates strategic leverage simultaneously creates ecological fragility.

Strategic-Military Geometry. Straits compress military space. They force naval forces into predictable trajectories, concentrate surveillance, and reduce the maneuvering room available for escalation management. The strategic-military subsystem examines the geometry of force projection: missile envelopes, air-defense arcs, submarine transit routes, chokepoint interdiction capabilities, and the spatial logic of deterrence. It also considers the role of alliances, basing arrangements, and regional security architectures.

A strait's military geometry is not defined solely by the states that border it; it is shaped by the broader strategic environment, including the presence of external powers and the distribution of advanced military technologies. In the Taiwan Strait, 2025 People's Liberation Army drills and expanded anti-access/area-denial (A2/AD) capabilities have tightened the operational envelope, compressing response windows for potential conflict to days rather than weeks.

Economic and Bio-Industrial Flows. Straits are conduits for the world's metabolism. Energy shipments, containerized goods, agricultural commodities, fertilizers, rare earths, and manufactured components all pass through these narrow corridors. The economic subsystem analyzes the density, composition, and directionality of these flows. It also examines the degree of dependency that states and industries have on specific straits, the elasticity of supply chains, and the availability of alternative routes.

The inclusion of bio-industrial flows—protein, grain, aquaculture inputs, and fertilizer—extends the analysis beyond traditional trade metrics. It reveals how straits underpin food security and the biological foundations of modern economies. The Strait of Malacca, for instance, carried 23.2 million barrels per day of oil in the first half of 2025—29 percent of global maritime oil trade—while simultaneously handling critical fertilizer and grain movements that sustain food systems across East and South Asia.

Labor and operational capacity. Chokepoint stability also depends on the human capacity to move ships and operate ports. Crews, pilots, tug operators, terminal labor, and maintenance specialists form an often-invisible throughput constraint: a corridor can be physically open and legally navigable, yet functionally degraded if qualified labor is unavailable or refuses to operate under perceived extreme risk.

During the Red Sea crisis, labor protections and bargaining frameworks reinforced seafarers' practical right to decline high-risk transits and demand repatriation, forcing reroutes irrespective of state preference. In SSA terms, labor behaves like a flow-governing variable inside the economic and bio-industrial subsystem, and its withdrawal can produce effects structurally similar to insurance withdrawal: a form of non-kinetic, threshold-driven closure that manifests first as temporal failure and can cascade into functional failure as schedules, port windows, and fleet capacity break down.

Bio-industrial governance (fisheries and aquaculture across borders). Bio-industrial flows are not only cargos; they are also living stocks governed by contested authority. Cross-border fisheries and expanding aquaculture corridors depend on licensing regimes, quota allocations, monitoring and enforcement capacity, and shared scientific baselines—often across states with conflicting sovereignty claims or unequal capabilities.

When governance is fragmented, the strait becomes a gray-zone interface: illegal, unreported, and unregulated (IUU) fishing, contested boarding and detention practices, and politically charged access disputes can generate escalation even without disruption to commercial shipping. Climate-driven shifts in fish distribution intensify this dynamic by changing what is at stake. As stocks move across boundaries, historic quota arrangements and enforcement patterns can become illegible, creating incentives for opportunistic harvesting and coercive signaling.

In SSA terms, ecological migration becomes a feedback loop: geophysical change reshapes bio-industrial flows, which stresses governance regimes and pushes enforcement into the civilizational and strategic-military subsystems through coast guard encounters and "white hull" assertion. Resilience in strait systems therefore requires not only pollution control but bio-industrial governance: shared monitoring, agreed rules for stock change, and credible enforcement that prevents fisheries conflict from becoming a parallel chokepoint crisis.

Governance Regimes. Straits are legal spaces as much as physical ones. Their governance is shaped by international conventions, historical treaties, domestic legislation, and contested territorial claims. The governance subsystem examines the legal status of each strait—whether it is an international strait, territorial sea, archipelagic passage, or special-regime zone—and the implications of that status for navigation rights, military transit, and resource exploitation. It also considers the role of institutions, dispute-resolution mechanisms, and the capacity of bordering states to enforce their claims.

Governance determines the rules of interaction; it defines what is permissible, what is contested, and what is strategically ambiguous. The Montreux Convention of 1936 continues to regulate warship transits through the Bosporus and Dardanelles, granting Turkey structural leverage that has been actively exercised during the Black Sea crises of the 2020s.

Demographic and Civilizational Interfaces. Straits are contact zones. They sit at the intersection of cultures, religions, languages, and demographic flows. The demographic-civilizational subsystem examines migration patterns, population pressures, cultural boundaries, and historical interactions. It reveals how straits function as gateways between civilizational spheres and how these interactions shape political identities, economic networks, and regional stability.

This dimension is often overlooked in strategic analysis, yet it is essential for understanding the long-term dynamics of straits. Civilizational interfaces influence governance, conflict potential, and the social foundations of maritime security. The Bab el-Mandeb, for example, remains a primary migration corridor between the Horn of Africa and the Arabian Peninsula, with flows driven by conflict, climate stress, and economic disparity that directly shape littoral governance and external basing strategies.

Technological Infrastructures. The final subsystem captures the technological forces reshaping maritime space. Autonomous shipping, satellite-based surveillance, undersea cables, cyber-nautical systems, and AI-driven logistics networks are transforming the operational environment of straits. The technological subsystem examines how these infrastructures enhance or undermine chokepoint resilience, how they alter the balance between offense and defense, and how they create new forms of vulnerability.

Technology does not eliminate the strategic significance of straits; it reconfigures it. It introduces new dependencies, new risks, and new opportunities for leverage. Undersea fiber-optic cables, for instance, converge heavily in the Strait of Malacca, the Strait of Hormuz, and the Bab el-Mandeb region; a single cable severance—whether accidental or deliberate—could disrupt global data traffic carrying trillions of dollars in daily financial transactions.

Over the longer term, the global energy transition is beginning to reshape technological and economic subsystems alike, gradually reducing dependence on traditional fossil-fuel corridors while creating new vulnerabilities in digital and renewable-energy infrastructure.

Interdependence and Systemic Dynamics

These six subsystems form the analytical core of the SSA model. They are not discrete categories but interacting components of a single architecture. A change in one subsystem reverberates through the others. A shift in geophysical conditions (Arctic ice melt) alters military geometry (new Northern Fleet access). A new governance regime reshapes economic flows (Montreux enforcement). A technological innovation modifies the balance of power (cyber-nautical disruption).

- Geophysical determinants establish baseline parameters for activity through physical constraints and their evolution (for example, the Bering Strait's extended navigable season).
- Strategic-military geometry defines escalation thresholds and leverage (for example, intensified A2/AD dynamics in the Taiwan Strait).
- Economic and bio-industrial flows determine systemic dependency and shock potential (for example, Malacca's energy throughput alongside food-system inputs).

- Governance regimes set rules of access and contestation (for example, Montreux constraints shaping warship transits in the Turkish Straits).
- Demographic and civilizational interfaces shape long-term stability and political narratives (for example, Bab el-Mandeb as a migration corridor influencing security and basing).
- Technological infrastructures introduce hybrid vulnerabilities and new leverage (for example, cable convergence and cyber-nautical disruption risks).

The SSA model captures these interactions, allowing for a holistic understanding of straits as dynamic systems embedded within the global order. Applied to any strait, it reveals unique configurations: the Taiwan Strait combines extreme military compression with unparalleled technological and civilizational stakes; the Strait of Hormuz layers geophysical narrowing with massive energy dependency and contested governance. No single subsystem dominates; their interplay determines whether a strait functions as a stable corridor, a volatile fulcrum, or a systemic risk node.

When primary chokepoints are disrupted, bypass geographies absorb massive systemic strain, redistributing pressure across secondary routes such as the Cape of Good Hope, the Northern Sea Route, and emerging corridors like the India–Middle East–Europe Corridor.

Applying the SSA Framework

This chapter establishes the framework that will be applied throughout the book. Each subsequent chapter examines a major strait through the lens of the SSA model, revealing how its unique configuration of subsystems shapes regional and global dynamics. By analyzing straits as systems rather than isolated features, the book provides a deeper understanding of the structural forces that govern the world's narrow places—and the vulnerabilities and opportunities they create.

In Hormuz, we will see extreme geophysical compression layered with massive energy dependency and low optionality. In the Bering Strait, we will observe a rapidly evolving geophysical subsystem reshaping strategic-military geometry. In the Taiwan Strait, we will encounter the convergence of military, technological, and civilizational stakes at their highest intensity.

The SSA model thus serves as both an analytical lens and a strategic compass for understanding the evolving architecture of global power.

4

Chapter 4

Straits as Civilizational Interfaces

Straits are not only strategic corridors; they are civilizational thresholds. They sit at the boundaries where languages, religions, economic systems, and political orders meet. Across history, these narrow maritime spaces have served as points of contact, exchange, and confrontation between societies. Geography forces proximity, and proximity produces interaction—trade, migration, cultural diffusion, or conflict. In every case, the strait becomes a site where civilizational trajectories intersect and where the structure of regional order is negotiated.

The civilizational significance of straits emerges from their dual nature as both connectors and separators. A strait links two bodies of water, enabling movement and exchange, yet it also divides two landmasses, marking the boundary between distinct cultural or political spheres. This duality creates a unique spatial tension. The Bosporus connects the Black Sea to the

Mediterranean while separating Europe from Asia. The Strait of Gibraltar links the Atlantic and Mediterranean while dividing the Iberian Peninsula from North Africa. The Strait of Malacca connects the Indian and Pacific Oceans while marking the interface between the Malay world and the broader Indo-Chinese sphere. In each case, the strait is both bridge and boundary.

As connectors, straits facilitate the movement of goods, ideas, technologies, and populations. They enable trade networks, cultural diffusion, and the formation of hybrid identities. As boundaries, they delineate political authority, cultural identity, and religious affiliation. They become sites where states project sovereignty, where empires confront one another, and where competing civilizational narratives collide. The tension between connection and separation is what gives straits their civilizational density.

Demographic Pressures and Identity Formation

The demographic dimension of straits reinforces this density. Populations near straits tend to be diverse, mobile, and economically active. Ports, trading hubs, and coastal cities emerge along these corridors, attracting merchants, migrants, and intermediaries. These populations often develop identities distinct from inland societies—more cosmopolitan, more hybrid, and more attuned to the rhythms of maritime exchange.

The cultural landscapes of Istanbul, Singapore, Tangier, and Aden illustrate how straits generate urban forms that reflect their role as civilizational interfaces. Istanbul's identity is shaped by centuries of Ottoman synthesis across Europe and Asia. Singapore's demographic composition—75.5 percent Chinese, 15.1 percent Malay, 7.6 percent Indian as of mid-2025, with roughly 19 percent inter-ethnic citizen marriages—embodies the hybridization produced by the Strait of Malacca's long-standing corridor function.

Migration flows intensify these dynamics. In 2025, the Eastern Route across the Bab el-Mandeb recorded approximately 238,000 movements in the first half of the year—an increase of 34 percent from 2024—driven by economic desperation, conflict, and climate stress from the Horn of Africa toward the Arabian Peninsula. These flows create social networks that span the corridor, but they also generate tensions within coastal communities and shape external powers' basing and security strategies. Climate-driven ecological stress amplifies these movements, demonstrating how geophysical change reverberates through the civilizational subsystem.

Engineered straits can produce even more dramatic demographic transformations. The construction of the Panama Canal brought tens of thousands of Afro-Caribbean laborers—primarily from Jamaica and Barbados—to the isthmus, permanently altering Panama's racial and cultural fabric. Unlike natural straits that evolve over centuries, engineered corridors can reshape the civilizational composition of an entire nation within a single generation.

The Arctic offers another profound civilizational interface. The Bering Strait has long been home to Yupik, Inupiat, and Chukchi communities whose familial, linguistic, and cultural ties have spanned the waterway for millennia. During the Cold War, the "Ice Curtain" severed these ancient connections, dividing families and communities for decades. Today, accelerating climate change and the retreat of Arctic sea ice are again transforming this interface, reopening possibilities for cross-strait interaction while introducing new geopolitical, environmental, and cultural pressures on Indigenous populations.

Straits also shape the political identities of the states that border them. Control over a strait often becomes central to national narratives, strategic doctrines, and historical memory. The Ottoman Empire's identity was inseparable from its control of the Bosporus and Dardanelles; contemporary Turkish foreign policy continues to treat the straits as a civilizational marker of strategic autonomy, leveraging the Montreux Convention to balance NATO, Russia, and regional ambitions. Iran's geopolitical posture is shaped by its position along the Strait of Hormuz, where Persian–Arab cultural and sectarian divides underpin narratives of sovereignty and resistance. China's strategic imagination is conditioned by the Taiwan Strait, where diverging identities have hardened: "Taiwanese-only" identification rose from 17.6 percent in the early 1990s to roughly 63 percent by 2023, with 2025 surveys indicating 76 percent or higher among younger cohorts. These identities influence foreign policy, alliance structures, and the willingness of states to escalate in defense of perceived core interests. Civilizational interfaces become strategic commitments.

Synthesis, Fault Lines, and Strategic Implications

Civilizational interaction within straits is not always peaceful. The same proximity that enables exchange also enables conflict. Straits have been sites of naval battles, sieges, blockades, and imperial contests. They have served as invasion routes and defensive barriers. The civilizational boundaries they mark can harden into fault lines, producing zones of friction where competing political orders confront one another.

The Taiwan Strait is a contemporary example of such a fault

line, where divergent political systems and national identities face each other across a narrow channel. The Bosporus has played this role repeatedly across centuries of imperial competition. In the Bab el-Mandeb, migration routes intersect with sectarian and cultural divides, contributing to the militarization of migration in which displaced populations are exploited as labor, leverage, or combatants within regional conflicts.

Yet conflict is only one expression of civilizational interaction. Straits also produce synthesis. They generate hybrid cultures, mixed economies, and shared institutions. They create spaces where differences are negotiated rather than enforced. The Mediterranean world, shaped by the interplay of its straits, is a historical example of such synthesis. The Indian Ocean littoral, structured by the Malacca and Sunda Straits, is another. Singapore's position at the Strait of Malacca exemplifies this ongoing synthesis: its hybrid demographic composition and inter-ethnic marriage rates demonstrate how maritime corridors can foster resilient cosmopolitan identities even amid great-power competition.

Colonial and post-colonial legacies continue to shape these dynamics. In Gibraltar, British imperial narratives clash with Spanish nationalist claims and Moroccan assertions, producing layered identity politics that complicate European Union and NATO positioning. In the Malacca region, post-colonial nation-building in Malaysia, Indonesia, and Singapore has been shaped by the need to navigate the historical imprint of Portuguese, Dutch, and British maritime empires while asserting modern sovereignty over one of the world's most vital trade corridors. These frictions show that civilizational interfaces are not only ancient; they are continually renegotiated through historical memory and contemporary power.

The Demographic-Civilizational Subsystem in the SSA Model

The demographic-civilizational subsystem of the Straits Systems Architecture captures these dynamics. It examines how cultural boundaries, demographic flows, historical interactions, and identity formations shape the strategic behavior of states and the long-term evolution of regional orders.

Across major straits, this subsystem manifests in distinct configurations:

- **Bab el-Mandeb** as an Arab–Horn of Africa interface, where migration becomes a vector for instability and external basing incentives.
- **Gibraltar** as a symbolic Europe–North Africa boundary shaping migration politics, alliance posture, and layered sovereignty narratives.
- **The Bosporus** as a Europe–Asia synthesis zone, granting Turkey leverage in NATO–Russia relations via the Montreux Convention.
- **The Taiwan Strait** as a hardened identity fault line with elevated great-power stakes.
- **The Strait of Malacca** as a Malay–Chinese–Indian interface, exemplified by Singapore's cosmopolitan demography and sensitive port-city system.

This subsystem interacts with the other five in the SSA model. Demographic pressures can alter governance regimes as migration reshapes littoral regulatory authority and internal political priorities. Civilizational narratives can intensify military geometry by raising identity-driven escalation risks or reshape economic flows through trust networks, boycotts, and diasporic commercial channels. A change in one dimension reverberates through the others, determining whether a strait functions as a zone of synthesis or a hardened fault line.

5

Chapter 5

Straits in the Age of AI and Autonomous Maritime Systems

The strategic significance of straits is being redefined by the rise of autonomous maritime systems, AI-driven logistics, and pervasive digital surveillance. These technologies do not diminish the importance of narrow maritime corridors; they transform the mechanisms through which power is exercised within them. The chokepoint of the twentieth century was shaped by physical interdiction, naval presence, and the geometry of sea lanes. The chokepoint of the twenty-first century is increasingly shaped by data, automation, and the ability to sense, predict, and influence movement at scale. Straits are becoming not only physical bottlenecks but digital ones.

The Five Technological Transformations

1. Autonomous Shipping

Maritime autonomous surface ships (MASS) and uncrewed surface vessels (USVs) are transitioning from experimental trials to operational scaling. In 2026, the U.S. firm Saronic raised $1.75 billion to accelerate production of its Corsair, Mirage, and Marauder platforms, while AUKUS partners expanded joint development of robotic maritime systems following successful 2025 trials. The International Maritime Organization advanced a goal-based, non-mandatory MASS Code in 2025, with mandatory provisions scheduled for 2028.

Autonomous systems promise reduced crew risk, lower operating costs, and persistent presence. Yet they introduce new dependencies: high-integrity navigation data, resilient satellite communications, and predictable traffic patterns. In compressed straits, where deviation is limited, a disruption in connectivity or algorithmic decision-making can immobilize fleets far more rapidly than traditional blockades. Algorithmic navigation errors in high-density corridors also amplify ecological risks, increasing the potential for collisions and environmental incidents.

2. AI-Driven Maritime Domain Awareness

Advances in satellite imaging, synthetic aperture radar, machine learning, and multi-source data fusion now enable near-continuous monitoring of vessel traffic—even when AIS signals are spoofed or disabled. Commercial platforms such as Planet's maritime domain awareness tools and Windward's AI systems detect "dark fleet" vessels and behavioral anomalies in real time across corridors like the Strait of Malacca and the Persian Gulf.

This transparency compresses escalation windows and elevates informational dominance. States that control or deny access to these surveillance layers gain asymmetric leverage over all actors transiting the strait. Increasingly, this dominance extends into the electromagnetic spectrum itself: AI-enabled fusion of RF data, radar returns, and optical imagery allows real-time detection and classification of emissions in littoral waters. In highly compressed straits, spectrum dominance has become as consequential as visual dominance.

3. Cyber-Nautical Vulnerabilities

As vessels, ports, and logistics systems digitize, straits become prime targets for non-kinetic disruption. GPS spoofing and jamming surged in contested corridors during 2025–2026: the Strait of Hormuz recorded hundreds of GNSS interference events, with vessels appearing on land or losing positional awareness entirely. Similar incidents in the Red Sea during the 2023–2025 crisis forced reliance on visual navigation and increased operational hazards.

Cyber operations can replicate the effects of a physical blockade—delaying traffic, corrupting manifests, or disabling port infrastructure—without visible naval forces. Attribution remains ambiguous, escalation ladders opaque, and the boundary between state and non-state actors blurred.

Major ports anchoring chokepoints, such as Singapore and Algeciras, are increasingly vulnerable through their "smart port" operational technology (OT) systems. Automated cranes, gantries, and terminal logistics controlled by OT networks can be compromised by ransomware or insider threats, halting cargo movement and creating physical backlogs in a strait without a single vessel being kinetically attacked.

4. Undersea Digital Infrastructure and Seabed Militarization

More than 1.4 million kilometers of submarine fiber-optic cables carry over 99 percent of global internet and financial data traffic, with major clusters converging in narrow corridors such as the Strait of Hormuz, the Strait of Malacca, the Bab el-Mandeb, and the Red Sea approaches to Suez. Repeated outages in the Red Sea in September 2025 and recurring disruptions in the Malacca region demonstrate the fragility: a single cable severance—whether from anchoring accidents, sabotage, or natural causes—can cascade into regional communications blackouts and financial instability.

Protection of this infrastructure is now central to chokepoint security. This vulnerability is compounded by the rapid militarization of the seabed. Governments are investing heavily in unmanned underwater vehicles (UUVs) for intelligence collection, mine countermeasures, and anti-submarine warfare. In gray-zone operations, these systems can also monitor or potentially sabotage undersea cables, turning the seabed into a contested domain layered beneath traditional chokepoints.

5. Algorithmic Logistics and Predictive Routing

AI systems now optimize global shipping routes in real time based on weather, geopolitical risk, insurance pricing, and port congestion. During the Red Sea crisis, these algorithms rapidly rerouted traffic around the Cape of Good Hope, demonstrating how digital systems can dynamically alter dependency structures. A strait that appears indispensable under one set of conditions may become bypassable under another; conversely, peripheral corridors can gain centrality as patterns shift.

This creates a feedback loop in which technological systems reshape the strategic meaning of physical geography. Algorithmic rerouting also interacts with climate-driven disruptions, as extreme weather and ecological pressures increasingly force endogenous shifts in global trade flows.

Hybrid Vulnerabilities and Systemic Risk

These technological transformations do not diminish the importance of straits; they deepen it. They add new layers of dependency, new forms of vulnerability, and new mechanisms of control. The chokepoint becomes a hybrid space where physical, digital, and informational systems intersect. The state or coalition that integrates these systems—combining maritime presence with data dominance, cyber resilience, and autonomous capabilities—gains structural advantage in the governance of straits.

In 2025–2026, the five transformations manifest in ways that directly reconfigure chokepoint power:

- Autonomous shipping introduces dependencies on communications and navigation integrity, making digital disruption capable of producing blockade-like effects.
- AI-driven domain awareness creates persistent transparency that compresses escalation time and elevates informational dominance, including in the electromagnetic spectrum.

- Cyber-nautical vulnerabilities enable non-kinetic disruption with ambiguous attribution, turning smart ports into chokepoint failure points.
- Undersea digital infrastructure concentrates systemic risk in cable clusters, while seabed militarization introduces new gray-zone tools for surveillance and sabotage.
- Algorithmic logistics reshapes dependency and optionality through predictive routing, redistributing pressure across corridors in response to conflict, insurance repricing, and climate-driven disruption.

Reconfiguring Chokepoint Power

The technological subsystem of the Straits Systems Architecture captures these dynamics. It reveals how the strategic significance of straits is being reconfigured by automation, surveillance, cyber operations, and digital infrastructure. It shows that the future of chokepoint power will be determined not only by geography and military capability but by the ability to control the informational and technological systems that sustain maritime movement.

Changes in this subsystem ripple through the other five: geophysical constraints shape where cables must be laid; governance regimes struggle to regulate autonomous transits; civilizational interfaces influence narratives around technological sovereignty.

This chapter completes the conceptual foundation of the SSA model. The chapters that follow apply this integrated framework to the world's major straits, revealing how each narrow place functions as a complex system shaped by geophysical constraints, military geometry, economic flows, governance regimes, civilizational interfaces, and technological infrastructures.

In Hormuz, we will see how cyber-nautical vulnerabilities and seabed warfare compound geophysical compression and energy dependency. In Malacca, we will observe how algorithmic logistics, cable concentration, smart-port OT risks, and AI-enabled spectrum dominance reshape systemic dependency and hybrid risk.

The analysis now shifts from theory to application—from the architecture of straits to the straits that structure the world.

6

Chapter 6

The Strait of Hormuz: Energy's Fulcrum

The Strait of Hormuz is the most strategically consequential energy corridor in the world. Its geometry is narrow, its traffic dense, and its systemic importance unmatched. No other strait channels such a high concentration of a single critical commodity—crude oil and petroleum products—through such a confined maritime space. In the first half of 2025, flows averaged 20.9 million barrels per day, roughly 20 percent of global petroleum liquids consumption and about one-quarter of all seaborne traded oil. The events of early 2026 reinforced this reality: Iran's effective closure of the strait in late February triggered production shut-ins peaking at 9.1 million b/d in April, pushing global markets into deficit and demonstrating how deeply the chokepoint anchors the international system.

CHAPTER 6

Geophysical and Military Compression

Although bordered by Iran and Oman, the Strait of Hormuz functions as an international waterway—a legally and strategically contested passage through which global shipping expects continuity of access. At its narrowest point, the strait is roughly 33 kilometers wide, yet navigable movement is funneled into tightly regulated traffic-separation lanes. Geographic width becomes operational compression.

Historically a conduit for high-value exchange across the Indian Ocean world, the corridor is now a principal artery for supertankers carrying crude oil and liquefied natural gas from Gulf producers—including Saudi Arabia, Kuwait, Iraq, Qatar, Bahrain, the UAE, and Iran—into external markets, most of them in Asia. Limited pipeline bypass capacity exists, but the majority of seaborne transit has no comparable alternative route. As a result, even the threat of disruption repeatedly triggers price spikes and systemic stress.

The geophysical determinants of Hormuz establish the foundation of its strategic significance. While the strait spans tens of kilometers, safe navigation is constrained to narrow lanes with minimal room for maneuver. Shallow waters restrict tanker movement and limit alternative routing. The curvature of the strait and its proximity to land create predictable vessel trajectories, increasing exposure to surveillance, interdiction, and attack. These physical constraints compress maritime movement into a corridor that is inherently vulnerable—precisely the condition that amplifies every other subsystem in the SSA model.

Geophysical compression also concentrates ecological externalities: heightened spill risk, concentrated emissions from idling vessels, and intensified marine disruption in a cluttered littoral environment.

Economic Flows, Dependency, and the Limits of Optionality

The economic and bio-industrial flows passing through Hormuz are central to its global significance. Beyond the 20.9 million b/d of oil and products in H1 2025, the corridor carries significant volumes of LNG and petrochemical feedstocks that underpin downstream industries. It also moves critical non-pipelineable commodities—refined products, chemicals, fertilizers, and industrial cargos whose transport depends on maritime lift.

The Gulf is a foundational supplier of key manufacturing inputs: roughly 50 percent of global seaborne sulfur, 23 percent of ammonia, 33 percent of helium, and 9 percent of aluminum. These flows sustain the energy metabolism of East Asia, South Asia, and parts of Europe.

The dependency structure is starkly asymmetric. Gulf producers possess some bypass capacity—Saudi Arabia's East–West Pipeline (up to 7 million b/d, with practical limits), the UAE's ADCOP line to Fujairah (about 1.5 million b/d), and Iran's Gorreh–Jask line (about 0.3 million b/d). But even at emergency utilization, these alternatives cannot replicate normal seaborne throughput. Terminal bottlenecks, storage constraints, and exposure to political risk further limit the speed at which optionality can be engineered.

Consumers—especially China, India, Japan, and South Korea—have few immediate substitutes. The 2026 closure and associated shut-ins of 7.5–9.1 million b/d produced immediate price spikes, inventory draws, and cascading industrial effects.

This trade basket forms a three-tier substitution structure:

1. **Partially bypassable crude** via finite pipelines, shifting exposure from sea lanes to fixed terrestrial infrastructure.
2. **Partially reroutable liquid streams**—refined products and some chemicals—at higher cost and risk, with limited terminal flexibility.
3. **Non-substitutable maritime cargos**—fertilizers, bulk agricultural inputs, containerized industrial goods, sulfur, ammonia, helium, aluminum—whose movement depends on shipping. Here, the chokepoint is not only a route constraint but a modality constraint.

A critical bio-industrial vulnerability compounds this exposure: the region's near-total dependence on desalination. Qatar, Kuwait, and the UAE derive most of their drinking water from coastal desalination plants. An oil spill or sustained disruption in the compressed waters of Hormuz could rapidly clog water intakes, threatening millions within days. Strikes on facilities during the 2026 crisis underscored how directly this infrastructure lies in the crosshairs.

Iran's export geography reinforces this fragility. Roughly 90 percent of Iranian crude exports originate at Kharg Island inside the Gulf, meaning that even with bypass infrastructure, the practical baseline remains an inside-the-Gulf loading architecture dependent on prevailing rules of confrontation and restraint.

Chokepoint coupling further compounds the problem. Crude diverted to Red Sea terminals must still traverse the Bab el-Mandeb, binding Gulf "escape routes" to Red Sea instability and creating a dual-chokepoint trap in which the global energy system can be constrained at both its Gulf exit and its southern gateway.

Bypass geographies redistribute rather than eliminate systemic strain.

Governance, Civilizational Interfaces, and Hybrid Risks

The governance regime surrounding Hormuz is shaped by contested interpretations of international law, historical grievances, and competing security architectures. Iran asserts regulatory authority and has repeatedly threatened closure. The United States and partners insist on transit passage under UNCLOS. This legal ambiguity creates a permanent zone of strategic friction. The absence of a shared governance framework increases the risk of miscalculation, as seen in the 2026 escalation when Iran imposed selective passage conditions and direct interdictions.

This governance deficit also explains the recurring cycle in which megaprojects to bypass Hormuz—canals, rail corridors—resurface during crises and fade afterward. In practice, incremental pipelines and terminal expansions have been favored because they deliver partial bypass capacity on shorter timelines, at lower cost, and under clearer sovereign control—while still preserving chokepoint coupling.

Demographic and civilizational interfaces add another layer of complexity. The Persian Gulf region is characterized by diverse populations, sectarian divides, and historical rivalries between Persian and Arab spheres. Hormuz sits at the intersection of these identities, shaping Iranian narratives of sovereignty and resistance while reinforcing Gulf Arab alignment with external security guarantors. These dynamics influence alliance structures, escalation thresholds, and the symbolic weight of incidents in the waterway.

Technological infrastructures further intensify hybrid risk. Undersea fiber-optic cables—at least seven major systems—converge in the corridor, carrying most Gulf data traffic. GPS spoofing and jamming have surged during contested periods, creating navigation hazards in an already compressed environment. Autonomous systems, satellite surveillance, and cyber-nautical networks add new layers of dependency and vulnerability.

Seabed militarization compounds this risk. UUVs are increasingly deployed for intelligence collection, mine countermeasures, and potential sabotage of undersea infrastructure. Major ports anchoring the chokepoint are exposed through "smart port" OT systems controlling automated cranes and terminal logistics; compromise of these systems can halt cargo movement without a single kinetic strike.

The 2026 crisis demonstrated how digital interference can compound physical restrictions, producing hybrid disruption at scale.

Hormuz as a Complete SSA System

The Strait of Hormuz exemplifies the full complexity of the Straits Systems Architecture. Its six subsystems interact to produce a uniquely dense and volatile configuration:

- **Geophysical determinants** create narrow lanes, predictable trajectories, and concentrated ecological risk.
- **Strategic-military geometry** pits Iranian asymmetric A2/AD capabilities against U.S. and coalition presence, embedding a multipolar deterrence paradox.
- **Economic and bio-industrial flows** include partially bypass-

able crude, limited reroutable liquids, and non-substitutable maritime cargos, alongside existential desalination dependence.

- **Governance regimes** feature contested transit-passage principles and exploitable legal ambiguity.
- **Demographic and civilizational interfaces** embed Persian–Arab identity and sectarian divides in escalation narratives.
- **Technological infrastructures** include cable clusters, GNSS interference, UUV activity, and OT vulnerabilities that expand coercive options.

Bypass geographies compound these dynamics by redistributing strain onto secondary routes such as the Red Sea corridor, itself exposed to Bab el-Mandeb risk.

Hormuz is not simply a narrow passage; it is a global fulcrum. Its stability underpins the functioning of the international system, and its disruption reveals the architecture of systemic risk.

The analysis now moves eastward to the Strait of Malacca, where trade volume, great-power competition, and technological integration converge in a different but equally consequential configuration.

7

Chapter 7

The Strait of Malacca: Asia's Artery

The Strait of Malacca is the primary maritime artery of the Indo-Pacific. It is the narrowest point in the world's most economically dynamic region, the hinge between the Indian and Pacific Oceans, and the corridor through which the strategic metabolism of East and Southeast Asia flows. In 2025, the strait recorded a new all-time high of 102,525 vessel transits—an 8.7 percent increase from 2024—making it the world's busiest commercial waterway. In the first half of 2025, it carried 23.2 million barrels per day of oil and petroleum products, equivalent to 29 percent of global maritime oil flows and surpassing even the Strait of Hormuz.

These flows—energy, containers, agricultural commodities, and digital infrastructure—make Malacca a structural component of the Asian system. Its stability underpins the functioning of the global economy. Its contemporary role is an extension of long historical continuity: Malacca has served for centuries as a primary gateway into and out of Asia, and its systemic importance expanded further as steamship networks and modern canals compressed distance between Europe and East Asia. Like all compressed corridors, it is a setting where small-scale disruption—criminal, accidental, or politically motivated—can generate outsized systemic effects.

The Duality of Density and Compression

Despite its centrality, Malacca's importance is often under-recognized outside maritime and security circles. Yet the strait's logic is unforgiving: because traffic density and schedule precision are high, a serious obstruction does not need to last long to produce global effects. The first-order outcome is immediate queueing and rerouting; the second-order outcome is a slowdown in trade velocity as ports, carriers, and manufacturers absorb delay into inventories, contract performance, and freight pricing. In a high-throughput corridor, even brief disruption becomes a macroeconomic event.

Malacca's systemic weight is amplified by the scale of energy and trade dependency it concentrates into a confined geometry. A substantial share of seaborne crude bound for East Asia transits the corridor, and an exceptionally large volume of global seaborne trade moves through the broader **Malacca–Sunda–Lombok** complex linking the Indian Ocean to the South China Sea. This concentration is enforced by physical compression: in key segments near Singapore, the navigable passage narrows to only a few kilometers, and at the tightest points it falls below 2 kilometers, converting a regional waterway into a global economic constraint.

The strategic implication is structural. As traffic density rises, the strait becomes a leverage node not only for littoral governance but for distant importers whose energy security and manufacturing continuity depend on a single observable corridor. Malacca also anchors the principal East–West maritime route that couples Indo-Pacific production networks to the Red Sea and Mediterranean system, making its stability inseparable from wider chokepoint dynamics.

Geophysical Determinants and Engineered Change

The geophysical determinants of Malacca define the constraints under which all other systems operate. The strait stretches roughly **900 kilometers**, with critical pinch points—especially the Philip Channel near Singapore—narrowing to **under 2.8 kilometers**, and in places **below 2 kilometers**. Bathymetry limits the passage of the largest tankers and deep-draft container ships, creating a practical "Malaccamax" constraint and pushing some traffic toward alternative routes such as the Lombok or Sunda Straits. Seasonal monsoons, sedimentation, and tidal dynamics further complicate navigation.

These natural processes are now compounded by aggressive human-driven geophysical change. Singapore's extensive land-reclamation projects alter coastlines, bathymetry, and marine ecosystems, degrading fisheries and triggering diplomatic friction—including periodic restrictions on sand exports by neighbors. Geophysical constraints in Malacca are therefore no longer purely natural; they are increasingly engineered and politically weaponized.

Compression concentrates ecological externalities: higher collision and grounding risk, elevated spill probability, and seasonal haze that can sharply reduce visibility and compound operational hazards.

Economic Flows, Dependency, and Costly Optionality

Malacca's narrow, shallow, high-traffic characteristics make it vulnerable to piracy, navigational accidents, and deliberate interdiction. When risk levels rise, shipping does not simply slow—it reconfigures. Diversion to the Sunda or Lombok Straits imposes distance and scheduling penalties: a Sunda diversion can add **~450 nautical miles** and **1–2 days**, while Lombok often adds several days. Even short diversions translate into multi-million-dollar daily cost increments at fleet scale through added fuel burn, charter time, schedule knock-on effects, and higher freight and insurance costs.

Risk in Malacca is not only geopolitical. Environmental and operational hazards—seasonal haze, dense cross-traffic, and cumulative collision probability—create a persistent background of disruption risk that scales with traffic volume. This has pushed governance toward a practical security-and-safety stack: traffic-separation schemes, enhanced vessel-traffic services, expanded navigational aids, and coordinated patrol architectures blending coast-guard presence with joint air-and-surface monitoring.

Because maritime density is so high, the stability of the strait depends as much on routine accident prevention and real-time traffic management as on crisis deterrence. Small incidents can cascade into queueing, insurance repricing, and schedule disruption across the Indo-Pacific logistics system.

Threat perceptions in Malacca have periodically been shaped by a convergence narrative: the concern that criminal tactics (piracy, opportunistic boarding) could be repurposed for politically motivated disruption in a corridor where even a single incident can generate outsized effects. Compression lowers the threshold at which low-probability events become high-consequence variables.

Strategic-Military Geometry: Distributed Deterrence

The strategic-military geometry of Malacca reflects the complexity of the Indo-Pacific security environment. The strait is bordered by Malaysia, Indonesia, and Singapore, but its significance extends far beyond them. China, Japan, South Korea, India, Australia, and the United States all depend on the corridor for energy and trade. This creates a multi-layered security architecture defined by interdependence rather than unilateral control.

No single actor dominates. Singapore's advanced surveillance capabilities, Indonesia's archipelagic geography, and Malaysia's coastal position distribute leverage. Recent U.S.–Indonesia defense cooperation and increased naval transits—such as U.S. amphibious groups rerouted during the 2026 Hormuz crisis—illustrate how extra-regional powers seek influence without direct sovereignty.

The narrowness of the strait limits surface maneuverability while elevating the value of air, submarine, and land-based systems for monitoring and denial. Governance disputes can function as strategic signals, shaping risk perceptions and external naval posture even without kinetic escalation.

For Beijing, the "Malacca Dilemma" is not only about the corridor's narrowness but about the surrounding alliance geometry. In a Taiwan contingency, Malacca can shift from background corridor to operational theater. Pressure is applied indirectly—through manipulation of energy and trade access that sustain warfighting capacity and economic endurance.

Economic and Bio-Industrial Flows: Coupled Vulnerabilities

In H1 2025, Malacca carried 23.2 million b/d of oil and petroleum products—29 percent of global maritime oil flows. China was the largest destination, receiving ~7.9 million b/d, or 48 percent of those volumes. Beyond energy, the corridor carries a substantial share of global container traffic linking East Asian manufacturing to markets across South Asia, the Middle East, Europe, and Africa.

Disruption propagates through two coupled channels:

- **Energy markets:** price shocks, futures repricing, inventory stress.
- **Container logistics:** time shocks, missed berthing windows, equipment imbalances, factory slowdowns.

Agricultural commodities, fertilizers, and protein flows move in both directions, binding food systems across continents. Optionality exists but is structurally expensive: Lombok and Sunda diversions cannot replicate Malacca's integrated logistics and informational architecture.

Thailand's proposed **Chumphon–Ranong Landbridge**, a ~$28 billion megaproject, illustrates the recurring chokepoint "threat cycle": ambitious bypass schemes surge during crises, generate political signaling, and recede when costs and geopolitical complications reassert themselves.

CHAPTER 7

Governance, Civilizational Interfaces, and Hybrid Risks

Governance in Malacca is shaped by national sovereignty, regional cooperation, and international norms. Malaysia, Indonesia, and Singapore share primary responsibility through mechanisms such as the Malacca Strait Patrols and the Aids to Navigation Fund. Singapore prioritizes freedom of navigation; Malaysia and Indonesia emphasize sovereignty and environmental protection. Littoral states reaffirmed their commitment to keeping the strait open under international law in April 2026.

Because Malacca is a shared corridor, proposals to monetize passage—such as tolling—carry strategic weight disproportionate to their fiscal return. Even suggesting tolls can generate intra-littoral friction and external scrutiny.

Civilizational interfaces add another layer. Malacca sits at the intersection of Malay, Chinese, Indian, and broader Southeast Asian cultural spheres. Singapore's hybrid demographic—75.5% Chinese, 15.1% Malay, 7.6% Indian, with high inter-ethnic marriage rates—reflects centuries of maritime synthesis. These interfaces shape political identities, economic networks, and the social foundations of maritime security.

Looking forward, Malacca's most plausible "infrastructure plan" is not a megaproject but an incremental hardening of the surveillance-and-response stack: fused AIS/radar/satellite/dron e feeds, shared information-fusion routines, and cyber-nautical defense of port OT systems and undersea cable infrastructure.

Malacca as a Complete SSA System

Malacca exemplifies a different configuration of the Straits Systems Architecture than Hormuz:

- **Geophysical determinants:** a 900-km corridor with pinch points under 2 km, Malaccamax constraints, and engineered bathymetric change.
- **Strategic-military geometry:** distributed deterrence among littoral states and major powers.
- **Economic and bio-industrial flows:** dense multi-commodity lift with costly optionality.
- **Governance regimes:** negotiated equilibrium of littoral stewardship under international norms.
- **Demographic-civilizational interfaces:** hybrid port-city ecosystems with deep historical synthesis.
- **Technological infrastructures:** cable hubs, autonomous-shipping initiatives, AI-enabled domain awareness, and cyber-nautical vulnerabilities.

The Strait of Malacca is a structural component of the Indo-Pacific system. Its stability is essential to the functioning of the global economy.

The analysis now moves westward to the **Bab el-Mandeb**, where energy, trade, and regional instability converge in a more volatile configuration.

8

Chapter 8

Bab el-Mandeb: The Gate of the Red Sea

The Bab el-Mandeb is one of the most strategically exposed straits in the global system. It is the southern gateway to the Red Sea and the Suez Canal—an engineered corridor that typically carries roughly 12 percent of global trade—and therefore a point where local coercion can scale into system-wide disruption. Its name—"the Gate of Tears"—captures the historical recognition that this narrow passage is both indispensable and vulnerable. Unlike the relatively stable governance structures surrounding the Strait of Malacca or the predictable geometry of the Strait of Hormuz, the Bab el-Mandeb sits at the intersection of fragile states, contested sovereignties, and overlapping security architectures. The Houthi attacks that began in late 2023 and persisted through 2025–early 2026 demonstrated this volatility in real time: Red Sea shipping traffic through the Bab el-Mandeb and Suez Canal fell by approximately 57–62 percent, Suez revenues collapsed by 40 percent or more in the early phases, and global rerouting around the Cape of Good Hope added 10–14 days to Asia–Europe voyages while driving freight rates sharply higher. The strait is a chokepoint defined by fragility and non-state leverage.

Whether Bab el-Mandeb is "more important" than Hormuz depends on the shock channel under analysis. Hormuz is the dominant single-commodity node: it concentrates a uniquely large share of global oil into one corridor, so disruption expresses itself first through energy-price repricing and immediate export constraints. Bab el-Mandeb is a higher network-leverage node: it gates the Red Sea–Suez chain, so sustained insecurity can strand an engineered corridor that normally carries a large share of interoceanic commerce and convert disruption into time—week-scale delays, charter repricing, inventory desynchronization, and inflationary pressure across multiple industries. In this sense, Hormuz is the archetypal energy fulcrum, while Bab el-Mandeb is a chokepoint-of-chokepoints whose systemic impact is mediated through trade velocity and the enforceability of transit passage.

Geophysical Compression and Regional Fragility

Geographically, the Bab el-Mandeb lies between Yemen on the Arabian Peninsula and Djibouti and Eritrea on the Horn of Africa. It connects the Red Sea to the Gulf of Aden and therefore functions as the southern hinge of the Suez system, linking the Mediterranean corridor to the wider Indian Ocean. Its Arabic name—"Gate of Tears"—has long captured the corridor's association with danger and loss. At its narrowest points the passage compresses to roughly 18 miles (about 29 kilometers), converting regional geography into a chokepoint whose disruption can propagate rapidly into global trade and energy markets.

The geophysical determinants of the Bab el-Mandeb establish the constraints under which all other systems operate. The strait is narrow, measuring approximately 26–31 kilometers at its widest and narrowing further at critical points, with two channels divided by Perim Island (Mayyun). The eastern channel (Bab Iskender) is about 5 kilometers wide and shallower in sections; the western channel is deeper but still constrained. On the African side, the Seven Brothers island group and associated reefs add additional navigational complexity near the Djibouti approaches. The surrounding coastline is rugged, arid, and sparsely populated, limiting the development of robust maritime infrastructure. Seasonal winds, currents, and tidal dynamics complicate navigation, particularly for smaller vessels. These physical constraints concentrate ecological externalities: the narrow, shallow waters heighten the risk of spills, groundings, and marine disruption in an ecologically sensitive interface between the Red Sea and the Gulf of Aden, where fisheries and coastal livelihoods remain vulnerable to both natural incidents and conflict-related pollution. This vulnerability has been weaponized. In March 2024 the Houthis sank the MV *Rubymar*, sending 21,000 metric tons of toxic ammonium phosphate sulfate fertilizer to the seabed. In August 2024 they boarded the Greek-flagged *Sounion*, carrying roughly 1 million barrels of crude, and deliberately detonated explosives on its deck, creating an imminent catastrophic environmental hazard. In narrow straits, asymmetric actors do not merely disrupt trade; they can hold marine ecosystems and coastal bio-industrial livelihoods hostage. The physical geometry compresses maritime movement into a confined space that is inherently difficult to secure. Geography does not merely shape vulnerability; it amplifies it.

Asymmetric Threats, External Basing, and Systemic Leverage

The strategic-military geometry of the Bab el-Mandeb reflects the fragmentation of the surrounding region. Yemen, Eritrea, and Djibouti border the strait, but none possesses the capacity to unilaterally secure it. Yemen's civil war has enabled Houthi forces—backed by Iran—to exert asymmetric control over the Yemeni coastline, using drones, missiles, unmanned surface vessels, and small-boat swarms to interdict shipping. This non-state capability transformed the Houthis into de facto gatekeepers of a passage carrying roughly 12 percent of global trade. Eritrea's isolation and Djibouti's limited resources create a power vacuum filled by external actors. Djibouti now hosts the world's highest concentration of foreign military facilities: the United States (Camp Lemonnier), China (PLA Support Base), France, Japan, and Italy maintain bases within kilometers of one another and critical cable landing stations. Djibouti's relative political stability and willingness to host external forces—especially compared with Yemen, Somalia, and Eritrea—helps explain why basing has consolidated there rather than dispersing across the wider littoral. This basing density is not incidental; it is an institutionalized power-projection anchor for states seeking to monitor, deter, and—when necessary—intervene across the Red Sea–Gulf of Aden interface. Unlike Hormuz, where a state actor can attempt overt closure, Bab el-Mandeb disruption is more often produced by non-state coercion and ambiguous capability, which can be offset—though not eliminated—by coalition escorts, air defense, and freedom-of-navigation operations. This density of external presence reflects the corridor's importance but also introduces competing security logics and escalation risks. The strait's military geometry is defined by overlapping jurisdictions, asymmetric threats, and the persistent risk of hybrid disruption from non-state actors.

This geometry has been further complicated by new state-on-state friction. In January 2024 Ethiopia, a landlocked state, signed a controversial Memorandum of Understanding with the breakaway region of Somaliland. In exchange for Ethiopian recognition of Somaliland's statehood, Addis Ababa secured a 50-year lease on a 20-kilometer stretch of coastline near the port of Berbera to develop a commercial and naval base. The agreement alarmed neighboring coastal states, particularly Somalia and Egypt, and introduced a new layer of overlapping spheres of influence along the approaches to Bab el-Mandeb. The African littoral is actively fracturing, adding state-level contestation to an already crowded field of external military presences.

Operationally, Bab el-Mandeb's strategic meaning lies in how quickly localized coercion can force system-wide rerouting. Because through-traffic to Suez cannot bypass the southern gate, persistent threat at Bab el-Mandeb converts the canal from an engineered shortcut into a stranded asset for interoceanic shipping. The recent threat environment has also shifted from the older piracy paradigm toward hybrid attack profiles—missiles, drones, and occasional boarding operations—against which traditional shipboard hardening measures are of limited value. Commercial adaptation has therefore combined market and military instruments: tightened routing discipline, selective use of private security (including the normalization of armed private guards provided by Private Maritime Security Companies on some routes), altered signaling practices, and reliance on coalition escorts and air-defense coverage. In such environments, private security becomes a layer in the governance stack: its presence can be required by insurers, constrained by flag-state and port-state rules, and overridden by crew refusal, illustrating how commercial protection, liability, and labor interact in chokepoint resilience. The economic consequence is temporal. Cape rerouting adds week-scale delay and higher fuel burn, ties up vessel days, and drives charter and insurance repricing that transmits into consumer inflation through container freight rates and higher delivered energy costs; depending on routing, operators have faced voyage extensions on the order of roughly ten to two-plus weeks, with Red Sea routings measured in the mid-30-day range and Cape routings often in the low-40s or higher. Tactically, practices developed for piracy risk—physical barriers, water cannons, and other close-in defensive measures—do not address stand-off missile and drone threats, reinforcing dependence on early warning, specialist risk intelligence, and, in higher-threat periods, escorted transit.

Economic Flows, Governance Fragmentation, and Hybrid Risks

The economic and bio-industrial flows that pass through the Bab el-Mandeb are central to its global significance. The strait is the southern gate to the Suez corridor and therefore a structural hinge in the Asia–Europe trade route, historically carrying containerized goods, energy shipments, and agricultural commodities. In approximate global terms, on the order of 8–10 percent of world seaborne trade and about 9 percent of global oil shipments move through the Bab el-Mandeb on routes connecting the Persian Gulf, the Red Sea, and markets in Europe and North America; liquefied natural gas from the Middle East bound for Europe also traverses this passage. Pre-crisis oil and petroleum liquids transits averaged around 9.3 million barrels per day in 2023 before plummeting to approximately 4.1–4.2 million barrels per day in 2024–early 2025. The corridor also supports food imports into the Middle East and Gulf hydrocarbon exports to Europe. The dependency structure is asymmetric: Europe and parts of the Middle East rely heavily on the route, while Asia has alternative Pacific options. The 2023–2026 crisis forced widespread rerouting around the Cape of Good Hope, increasing fuel consumption by 30–50 percent, extending transit times, and producing cascading effects on global supply chains. The strait is therefore a structural component of the global economic system, and its instability introduces systemic risk.

To mitigate these disruptions, shipping companies established overland trucking "land bridges" from ports in the UAE and Bahrain across Saudi Arabia and Jordan to Mediterranean ports such as Haifa. These routes illustrate the modality constraints that define chokepoint resilience: while overland trucking can provide faster transit for high-value or time-sensitive containerized goods, it is physically impossible to substitute the massive bulk volumes of grain, fertilizer, or crude oil that a single large vessel carries.

The governance regime surrounding the Bab el-Mandeb is fragmented. Yemen's civil war has created a contested political environment in which multiple actors, including the Houthis, claim authority over coastal waters. Eritrea's governance is centralized but opaque, and Djibouti's regulatory capacity is limited. International law provides a framework for transit passage, but enforcement is inconsistent amid ongoing conflict; as an international strait, the corridor is not a lawful venue for unilateral tolling or tariffs on passage. Because the strait is the southern gate to Suez, sustained interdiction or closure at Bab el-Mandeb functionally severs the Suez corridor's connectivity for commercial shipping, undermining the premise of uninterrupted east–west passage. The absence of a coherent governance structure increases the risk of piracy, smuggling, and maritime insecurity. It also complicates efforts to establish stable rules of interaction among the external powers that operate in the region. Governance in the Bab el-Mandeb is not institutionalized; it is improvised and heavily dependent on ad hoc naval coalitions.

Legally, the Bab el-Mandeb is best understood as an international strait used for international navigation and therefore governed primarily by the transit-passage logic of UNCLOS: passage is to be continuous and expeditious, and coastal authority is constrained from measures that would have the effect of suspending through-traffic. This differs from innocent passage in territorial seas, where broader regulatory discretion exists. The practical consequence is systemic: because Suez is an engineered corridor embedded inside a longer legal and operational chain, disruption at Bab el-Mandeb does not merely slow shipping in the southern Red Sea—it collapses the functional value of Suez for interoceanic transit, forcing traffic into a different global geometry (Cape routing) even if the canal itself remains open. In this configuration, legal status and physical geography fuse: the stability of "east–west passage" depends on the enforceability of transit norms at the southern gate, and when enforcement fails, risk pricing, insurance conditions, and coalition naval posture become the de facto governance mechanism.

The demographic and civilizational interfaces surrounding the strait add another layer of complexity. The Bab el-Mandeb sits at the intersection of the Arab world, the Horn of Africa, and the broader Indian Ocean cultural sphere. Migration flows across the strait remain significant despite the security crisis. In 2025 the Eastern Route recorded approximately 506,600 tracked movements—an 18 percent increase from 2024—with arrivals in Yemen from the Horn of Africa rising sharply to 107,900. The route was the deadliest on record, with 922 deaths or missing persons reported (double the previous year). These flows, driven by economic opportunity, conflict, and environmental stress, create social networks that span the corridor but also generate tensions within coastal communities and fuel exploitation by armed groups. The civilizational dynamics of the region influence political identities, patterns of conflict, and the social foundations of maritime security. The strait is not only a strategic corridor; it is a human corridor shaped by mobility, displacement, and cultural exchange.

The technological infrastructures that support maritime movement through the Bab el-Mandeb are increasingly central to its strategic dynamics. Undersea cables—15 to 17 major systems—converge in the narrow corridor, carrying the bulk of data traffic between Europe, Asia, and the Middle East. Multiple cable cuts in 2024–2025 (including incidents in March 2024 linked to a sunken vessel and September 2025 near Jeddah) caused widespread internet disruptions and latency spikes in Asia and the Middle East. Cyber-nautical risks, data integrity issues, and the security of digital infrastructure are now integral components of chokepoint stability. This vulnerability is compounded by the rapid militarization of the seabed and the digitization of port operations. Unmanned Underwater Vehicles (UUVs) are increasingly relevant for intelligence collection and potential sabotage of undersea infrastructure in gray-zone operations, while major regional ports and cable landing stations face exposure through "smart port" operational technology (OT) systems controlling automated cranes, gantries, and terminal logistics. Compromise of these OT networks can create physical shipping backlogs without kinetic attacks on vessels. The digitization of maritime logistics introduces the possibility of non-kinetic disruption, particularly in a region where state capacity is uneven and external actors operate with divergent priorities. The chokepoint is no longer purely physical; it is also informational and hybrid.

The Bab el-Mandeb exemplifies a configuration of the Straits Systems Architecture defined by fragility. Its geophysical constraints, fragmented governance, asymmetric threats from non-state actors, dense external basing, high migration pressures, and concentrated undersea infrastructure create a strategic environment that is volatile and difficult to stabilize. In the Bab el-Mandeb as of 2025–2026, geophysical determinants create a narrow corridor (26–31 km at widest, with critical channels divided by Perim Island and additional complexity from the Seven Brothers group) that imposes high compression, limited maneuverability, and concentrated ecological risk in an environmentally sensitive interface—risks that have been weaponized through deliberate sinking and detonation of vessels carrying toxic fertilizer and crude oil; strategic-military geometry features Houthi asymmetric A2/AD capabilities (drones, missiles, unmanned surface vessels) layered against a dense concentration of foreign military facilities in Djibouti and new state-level contestation introduced by Ethiopia's 2024 MoU with Somaliland for a naval and commercial base near Berbera, producing non-state leverage, overlapping spheres of influence, escort dependence, and persistent hybrid escalation risk that external basing can mitigate but not eliminate; economic and bio-industrial flows encompass container and energy movements with limited slack, where route substitution via the Cape of Good Hope remains possible but imposes substantial cost and time penalties (freight, insurance, inventory stress, and week-scale delays that drive charter repricing and inflation), while overland land-bridge trucking offers partial relief for high-value containers but cannot substitute the bulk volumes of grain, fertilizer, or oil carried by large vessels, and digital-cable concentration and time-sensitive logistics create modality constraints beyond simple rerouting; governance regimes are marked by fragmented authority amid Yemen's civil war and contested enforcement of transit norms under UNCLOS, resulting in improvised security arrangements and elevated miscalculation risk; demographic and civilizational interfaces sustain an Arab–Horn of Africa interface characterized by high migration pressure (506,600 tracked movements on the Eastern Route in 2025 with 922 deaths or missing), functioning as both an instability vector and a vector for cultural exchange and exploitation; and technological infrastructures feature 15–17 undersea cables with repeated cuts and outages in 2024–2025, alongside emerging hybrid risks from UUV seabed operations, smart-port OT vulnerabilities, and expanding AI-enabled electronic warfare, generating layered physical-digital exposure that amplifies systemic risk.

The strait is not merely a narrow passage; it is a structural vulnerability in the global system. Its stability depends on a delicate balance of external intervention, regional cooperation, and the resilience of global supply chains. This chapter demonstrates how the SSA model reveals the internal logic of a major strait whose significance is rooted in volatility, fragmentation, and systemic exposure. The analysis now moves northward to the Suez Canal and its approaches, where the dynamics of engineered geography, global trade, and chokepoint risk converge in a different but equally consequential configuration.

9

Chapter 9

Suez: The Engineered Strait

The Suez Canal is the world's most consequential engineered strait. Unlike natural chokepoints shaped by geology and hydrology, Suez is a deliberate intervention in geography—a manufactured corridor that reconfigured global trade, altered the strategic geometry of three continents, and reshaped the economic metabolism of the modern world. Its significance does not arise from inherited compression but from the decision to impose a new form of compression on the landscape. Suez is a reminder that chokepoints are not only discovered; they can be constructed.

The canal's engineered character is inseparable from the political economy of its creation. Mid-nineteenth-century concessions mobilized cross-border capital and assembled a vast labor force—initially through coercive corvée labor, later through mechanized excavation. The project produced not only a channel but an entire settlement ecology: canal cities, logistics nodes, and informal economies emerged along the route. Suez also introduced a profound ecological externality by enabling large-scale species migration between the Red Sea and the Mediterranean—**Lessepsian migration**—which has transformed regional ecosystems. More than 400 alien species have crossed into the Mediterranean, with herbivorous rabbitfish overgrazing algal forests and converting them into barren rocky zones. Along the Eastern Mediterranean coast, Lessepsian migrants now constitute more than half of trawl catches, displacing native species and triggering irreversible food-web shifts.

Suez exemplifies how an engineered chokepoint is simultaneously an infrastructural system, a governance project, and an ecological intervention that permanently reshapes its environment.

Engineered Geography and Systemic Fragility

The geophysical determinants of Suez differ fundamentally from those of natural straits. The canal is a linear, shallow, tightly regulated waterway with no natural currents, no tidal complexity, and no alternative navigable channels. Its geometry is engineered for efficiency but constrained by depth (approximately 24 meters) and width, imposing strict requirements on vessel size, speed, and spacing. The 2014–2015 expansion added passing capacity and partial two-way segments, but the system remains a single engineered corridor whose failure modes are inherently non-redundant.

This predictability comes at the cost of inflexibility. When movement is disrupted—by accident, mechanical failure, or external threat—the entire system halts. The **2021 Ever Given** grounding demonstrated this fragility: a single vessel blocked the canal for six days, imposing global economic costs estimated at $9–10 billion per week.

The strategic-military geometry of Suez reflects its position at the intersection of multiple security architectures. The canal links the Mediterranean and Red Seas, connecting the maritime domains of Europe, the Middle East, and the Indian Ocean. Egypt exercises sovereign control, but the corridor is embedded in the interests of external powers— the United States, European states, Gulf countries, and increasingly China—whose military mobility and logistical reach depend on uninterrupted passage.

The **2023–2026 Red Sea crisis** amplified these dynamics. Houthi attacks on Bab el-Mandeb approaches forced widespread rerouting, exposing Suez to cascading risks even as Egypt maintained operations. Hybrid threats extend beyond the canal banks: undersea cable vulnerabilities, cyber-nautical manipulation, and interference with traffic-management systems create non-kinetic avenues for disruption.

Economic Flows, Substitution Tiers, and Strategic Exposure

Pre-crisis, Suez carried up to **15 percent of global trade**, including a disproportionate share of East–West container flows, energy shipments, and bulk commodities. The Red Sea crisis triggered a collapse: container transits fell by **86–90 percent** at peak, and overall traffic remained roughly **60 percent** below pre-crisis levels into early 2026. Revenues plunged from a 2023 record of ~$10.25 billion to ~$4 billion in 2024—a decline of more than 60 percent.

Partial recovery began in 2025, with the first half of FY 2025/2026 showing increases in vessel count (+5.8%), net tonnage (+16%), and revenue (+18.5%). By early 2026, the canal recorded 1,315 vessels, 56 million tons, and $449 million in revenue—up from the same period in 2025.

Suez's dependency structure is asymmetric. Europe relies heavily on the canal for access to Asian markets; Asia has costlier alternatives via the Cape of Good Hope. The canal also supports Middle Eastern food systems through grain, protein, and agricultural inputs. Suez is not merely a conduit; it is a structural component of global supply chains whose disruption produces immediate systemic effects.

The trade basket can be read as a **tiered substitution structure**:

1. **Time-sensitive container flows** Diversion is possible but expresses itself as inventory desynchronization, missed berthing windows, and contractual penalties.
2. **Bulk and energy cargos** Reroutable via the Cape, but at the cost of inflated ton-miles, fuel burn, and insurance premiums.
3. **Engineered single-channel exposure** The most severe failure mode is stoppage: one obstruction halts the entire corridor and triggers nonlinear global delay cascades.

Egypt's **SUMED pipeline** illustrates the limits of engineered optionality. With ~2.5 million b/d capacity, it allows crude to bypass the canal—but only if Bab el-Mandeb is secure. Because SUMED's input terminal lies in the Red Sea, insecurity at the southern gate neutralizes both the canal and its bypass infrastructure, creating a **dual-chokepoint trap**.

The collapse of canal revenues exposed Egypt's sovereign solvency. The 2024 revenue shock contributed to a sharp currency devaluation and forced Egypt to secure a major external support package, including IMF-linked assistance and the UAE's Ras el-Hekma development deal. When an engineered chokepoint fails, it can threaten not only global supply chains but the fiscal stability of the host state.

Governance, Historical Contingency, and Hybrid Risks

The Suez Canal Authority (SCA) manages operations under Egyptian sovereignty. The **1888 Convention of Constantinople** guarantees passage during peace and war, but enforcement depends on Egypt's political stability and strategic calculus. The canal's modern history underscores this contingency: nationalization in 1956 and closure from 1967 to 1975 demonstrated that legal guarantees do not eliminate the possibility of state-driven interruption.

The **1956 Suez Crisis** revealed how engineered chokepoints can reorder international systems. Egypt's nationalization triggered a British-French-Israeli intervention, but U.S. and Soviet pressure forced withdrawal. The episode marked the decline of European coercive capacity and the rise of the United States as the principal external guarantor of Western access to Middle Eastern energy and sea lines of communication.

In SSA terms, 1956 shows that chokepoint leverage can be exercised through **law, force, and finance** simultaneously—and that the credibility of "free passage" ultimately rests on the prevailing balance of power.

Demographic and Civilizational Interfaces

Suez sits at the boundary between Africa and Asia, between the Arab world and the Mediterranean sphere. Canal cities—Port Said, Ismailia, Suez—reflect a history of migration, labor flows, and cultural exchange shaped by construction and operation. The canal is embedded in Egyptian national identity, symbolizing sovereignty, modernization, and geopolitical relevance. This symbolic dimension influences political decisions regarding its management and security.

Technological Infrastructures and Hybrid Exposure

The canal relies on advanced dredging systems, traffic-management algorithms, satellite navigation, and digital logistics networks. Undersea cables converge in the Red Sea approaches, forming part of the broader Bab el-Mandeb/Red Sea cluster that carries Europe–Asia–Middle East data traffic. Cable cuts in 2024–2025 caused regional internet disruptions and latency spikes.

Cyber-nautical risks—including manipulation of navigation systems, manifests, or port operations—introduce new forms of chokepoint vulnerability. UUVs increase gray-zone risks to seabed infrastructure. Smart-port OT systems create additional attack surfaces. In an engineered corridor, technological failure can halt movement as effectively as a physical obstruction.

Comparative Lens: Suez and Panama

Suez and Panama demonstrate that engineered chokepoints share structural function but differ in operating physics and failure modes.

- **Suez** is obstruction-native: its primary vulnerability is geometric.
- **Panama** is climate-native: its primary vulnerability is hydrological.

Suez's failure mode is **functional stoppage**; Panama's is **temporal degradation** that cascades into partial functional failure. Both reveal that engineered chokepoints are not isolated points but nodes embedded in wider chains whose weak links may lie outside the corridor itself.

Suez as a Complete SSA System

Suez exemplifies a unique configuration of the Straits Systems Architecture:

- **Geophysical determinants:** linear engineered channel, depth/width constraints, single-point blockage risk, ecological transformation via Lessepsian migration.
- **Strategic-military geometry:** Egyptian sovereignty overlaid with external basing interests and Red Sea spillover.
- **Economic and bio-industrial flows:** containerized just-in-time flows, reroutable bulk/energy cargos, SUMED dependence on Bab el-Mandeb, sovereign solvency exposure.

- **Governance regimes:** centralized SCA management within a politically contingent framework.
- **Demographic-civilizational interfaces:** Africa–Asia boundary, canal-city identity, national symbolism.
- **Technological infrastructures:** cable convergence, cyber-nautical risk, UUV exposure, OT vulnerabilities.

Suez is an engineered strait whose significance is rooted in design, dependency, and systemic exposure. The analysis now moves westward to the **Strait of Gibraltar**, where Atlantic–Mediterranean exchange, great-power presence, and civilizational boundaries converge in a different but equally consequential configuration.

10

Chapter 10

Gibraltar: The Atlantic–Mediterranean Interface

The Strait of Gibraltar is the western gateway to the Mediterranean and the point where the Atlantic Ocean compresses into a narrow corridor before entering one of the world's most historically dense maritime basins. Its strategic significance derives from its position at the intersection of two maritime systems, two continents, and multiple civilizational spheres. Unlike the volatility of the Bab el-Mandeb or the engineered precision of Suez, Gibraltar's importance is rooted in continuity: it has been a strategic fulcrum for millennia, and its relevance persists in the contemporary global order. Each year, on the order of 100,000 vessels transit the strait, transporting a significant share of global maritime trade. In 2025, traffic through the Strait of Gibraltar from the ports of Algeciras and Tarifa was on track for record highs, exceeding 3.2 million passengers and 700,000 vehicles by July—underscoring Gibraltar's enduring role as a hinge between the Atlantic and Mediterranean worlds and a high-frequency Europe–Africa interface.

Gibraltar is also a place where chokepoint power is immediately visible. From the Rock's elevated batteries, the strait reads as a layered operating environment: dozens of vessels in transit or anchorage, ferry crossings stitching Europe to Africa, and naval platforms co-present with commercial traffic. This visibility is not incidental—it is a product of intersecting systems. Terrain and line-of-sight combine with port services, bunkering economics, radar and sensor coverage, and even local microclimates to produce a strategic location whose value is continuously re-made. Gibraltar's "unchanging" geography matters, but its strategic meaning is emergent: shifts in logistics networks, military technology, domestic politics, or surveillance architectures can quickly make the corridor feel more controllable—or less—without the coastline moving an inch.

This perspective helps avoid a false choice between "ideas" and "realities." Gibraltar's material geography sets hard constraints, but its geopolitical weight is not a fixed essence derived from rock and water alone. It is produced through the alignment of multiple subsystems: shipping networks and port services that concentrate traffic; legal regimes that define passage and contestation; sensors and data architectures that make movement legible; and environmental conditions that shape what can be seen, tracked, and intercepted. When these assemblages shift—through new military technologies, changes in logistics routing, or governance realignments—the same place can become more strategic, less strategic, or strategic in a different way. In Gibraltar, place is therefore not merely a container for geopolitics; it is an active interface continuously remade by the systems that pass through it.

CHAPTER 10

Geophysical Foundations and Emergent Strategic Meaning

The geophysical determinants of the Strait of Gibraltar establish the foundation of its strategic role. The strait is narrow—approximately 13–14 kilometers at its closest point—with strong currents driven by the salinity and temperature differentials between the denser Mediterranean waters and the lighter Atlantic inflow. Its depth varies significantly, ranging from several hundred meters in many sections to deeper basins that shape submarine movement and acoustic conditions. The physical geometry compresses maritime traffic into predictable surface corridors while simultaneously enabling deepwater transit for submarines and large vessels. This combination of surface constraint and subsurface depth makes Gibraltar a unique environment for both commercial and military navigation. Geography does not merely shape movement; it structures the strategic possibilities of the region. These constraints also concentrate ecological externalities and operational hazards. Since 2020, mariners and researchers have documented more than 500 reported interactions between a small, critically endangered Iberian orca subpopulation and boats in the wider Strait of Gibraltar/Iberian coastal region, with many incidents involving contact with rudders on small craft—adding a localized but persistent safety variable in an already high-frequency corridor.

Gibraltar's role as a gateway is not only historical but geological. In deep time, the stability of Atlantic–Mediterranean connectivity has hinged on whether a Gibraltar-scale corridor remains open, restricted, or closed. During the late Miocene—roughly 5–6 million years ago—the Mediterranean experienced the Messinian Salinity Crisis, when restriction of Atlantic inflow contributed to extreme evaporation, basin-wide salt deposition, and major reconfiguration of river systems draining into the Mediterranean. The subsequent reestablishment of Atlantic connection at the end of the crisis—often modeled as a rapid reflooding event—illustrates a foundational principle of chokepoint analysis: changes in the geometry of a gateway can reshape not only trade routes but entire basins, ecosystems, and coastlines. The evaporite record also implies that Atlantic water inputs persisted at least intermittently during parts of the crisis, even if outflow was restricted and the exact configuration of gateways remains debated. Gibraltar therefore exemplifies how a narrow passage can function, across vastly different timescales, as an on/off valve for system-wide transformation. Many reconstructions also imply that other Atlantic connections through southern Spain and northern Morocco progressively diminished before Gibraltar emerged as the primary gateway, though the exact sequencing is difficult to resolve because the final stages of closure are often erased by uplift and erosion.

The deeper lesson is that gateway history is reconstructed under uncertainty. In the late Miocene, Atlantic–Mediterranean exchange likely moved through a shifting network of passages across southern Spain and northern Morocco (often framed as Betic and Rifian corridors) and possibly through the Gibraltar region itself. Many onshore records suggest that the Moroccan (Rifian) connections were already being restricted and closing by the latest Tortonian into the earliest Messinian, with southern branches shutting earlier and northern branches likely short-lived thereafter. Precisely when each corridor became too shallow, intermittent, or fully closed is difficult to pin down because the act of closure is often accompanied by uplift and erosion that remove the very sediments that would record the final stages of connection. This creates an interpretive asymmetry: the absence of preserved marine deposits is not, by itself, proof that no marine gateway existed. For chokepoint analysis, the implication is methodological as well as geophysical: the geometry of a gateway is path-dependent, and the most consequential transitions can be the hardest to observe directly—yet they can still reorganize entire systems.

The strategic-military geometry of Gibraltar reflects its long history as a contested maritime gateway. The United Kingdom maintains control of the Rock of Gibraltar, providing commanding visibility over the strait and access to one of the most strategically located ports and air bases in the world. Spain contests aspects of this control, creating persistent political tension that overlays the military environment. Gibraltar's durability as a military outpost is rooted in physical defensibility: the Rock's steep profiles, layered fortifications, and tunneled infrastructure have historically enabled control to be held under siege and blockade conditions, turning geology into endurance. NATO's presence in the region, combined with the interests of European, North African, and transatlantic actors, creates a layered security architecture. The strait's narrowness limits surface maneuverability but provides opportunities for submarine transit, surveillance, and anti-submarine operations. Recent U.S. carrier strike group transits and increased Russian naval activity in the western Mediterranean have heightened the strait's operational tempo. Gibraltar's military geometry is defined by observation, access, and the ability to monitor and influence movement between two maritime domains.

Gibraltar also highlights how chokepoint control becomes three-dimensional. Submarine warfare and anti-submarine surveillance turned the strait's depth, thermohaline layering, and acoustic environment into strategic variables alongside surface traffic. During the Cold War, monitoring this corridor was treated as urgent because submarines transiting between the Mediterranean and Atlantic could exploit complex currents, sills, and water-mass boundaries to reduce detectability. In parallel, intensified oceanographic research at Gibraltar—often organized through transnational scientific and alliance frameworks—helped refine understanding of Mediterranean outflow and its role in wider Atlantic circulation. The structural lesson for the SSA model is that geophysical determinants and technological infrastructures co-evolve: the same measurements that improve scientific models of global circulation can also improve the surveillance architectures that govern passage.

Economic Flows, Transshipment, and Modality Constraints

The economic and bio-industrial flows that pass through the Strait of Gibraltar are central to its global significance. The strait is a major conduit for container traffic, energy shipments, and bulk commodities moving between Europe, Africa, and the broader Atlantic world. Tanger Med port on the Moroccan side has become a dominant transshipment hub, handling 11,106,164 TEUs in 2025 and anchoring a large share of western Mediterranean container interchange. The corridor also supports Spain's six regasification terminals and critical energy pipelines such as the Medgaz line connecting Algeria to Spain. Agricultural products, manufactured goods, and energy flows transit the corridor in both directions. The dependency structure is distributed: Europe relies on the strait for access to Atlantic markets and African resources, while North Africa depends on it for trade with Europe. In this interface role, diversion is possible but rarely neutral: it shifts burdens into transshipment capacity, LNG and pipeline operations, and the hybrid protection requirements of cables, terminals, and ports. The strait is not merely a passage; it is a structural component of the economic architecture of the western Mediterranean and the broader transatlantic system.

Gibraltar's trade logic is also geometric. As the only natural entrance to the Mediterranean, it is the mandatory western gate for any vessel using the Suez corridor to link Europe with the Red Sea, the Gulf, and the Indo-Pacific. When the Suez–Red Sea route is usable, the Gibraltar–Suez chain delivers major distance and time savings relative to Cape routing—on the order of several thousand nautical miles for typical Europe–Asia itineraries—translating into lower fuel burn, fewer vessel-days, and tighter schedule integrity for container networks. The inverse is equally important: when Gibraltar is disrupted, the Mediterranean becomes a semi-enclosed basin for through-traffic, and interoceanic routing collapses back toward longer Atlantic circuits; the result is a systemic time-and-cost shock that resembles, in structural form, the Ever Given–style logic of single-point failure.

Governance, Civilizational Interfaces, and Hybrid Risks

The governance regime of the Strait of Gibraltar is shaped by overlapping sovereignties, historical treaties, and international norms. The United Kingdom administers Gibraltar (anchored in the Treaty of Utrecht, 1713), Spain controls the northern shore, and Morocco governs the southern coast. The legal status of the strait under international law guarantees freedom of navigation, but sovereignty disputes and border management have long been sources of friction. In February 2026, the full text of a UK–EU treaty framework on Gibraltar was published that provides for the removal of the border fence and a Schengen-style arrangement: Spain would assume responsibility for Schengen controls at Gibraltar's airport and port, enabling the free movement of people and goods across the land frontier while preserving the parties' sovereignty positions. This negotiated governance architecture illustrates a core chokepoint reality: stability in narrow places is often produced through complex shared-administration arrangements rather than unilateral control.

The demographic and civilizational interfaces surrounding Gibraltar add depth to its strategic significance. The strait sits at the boundary between Europe and Africa, between the Christian and Islamic worlds, and between the Atlantic and Mediterranean cultural spheres. Its own naming encodes this layered history: Gibraltar derives from the Arabic *Jabal Ṭāriq* ("Mountain of Tariq"), while classical antiquity framed the passage as the "Pillars of Hercules." Migration pressure across the western Mediterranean route remains a persistent governance and security variable. In the first quarter of 2026, Frontex reported about 4,400 detections on the Western Mediterranean route, the only major route showing an increase year-on-year in that period. These flows shape political debates, enforcement posture, and humanitarian risk across both shores. Gibraltar is not only a strategic corridor; it is a civilizational threshold where mobility, identity, and security narratives intersect.

The technological infrastructures that support maritime movement through the Strait of Gibraltar are increasingly central to its strategic dynamics. Undersea cables and energy pipelines converge in the region, linking Europe and North Africa and forming part of the critical digital backbone for transatlantic connectivity. Advanced surveillance systems, maritime domain awareness platforms, and AI-driven logistics networks enhance the ability of states to monitor and manage traffic. Recommendations for enhanced cable protection—such as deeper burial and reinforced outer linings—highlight emerging hybrid risks. Cyber-nautical threats introduce new forms of chokepoint vulnerability, particularly in a region where multiple actors with divergent interests operate in close proximity. The strait is not only a physical corridor but a digital one, and control over its informational environment is becoming a form of strategic leverage.

The Strait of Gibraltar exemplifies a Straits Systems Architecture configuration defined by continuity, observation, and civilizational density. Its strategic significance derives from its position at the hinge between two maritime systems—the Atlantic and the Mediterranean—and at the meeting point of two continents. Its vulnerabilities emerge from overlapping sovereignties, persistent migration pressures, and the growing integration of digital infrastructures. As climate change, energy-transition routing, and shifting power dynamics reshape maritime risk, Gibraltar's role as a stable Atlantic–Mediterranean gateway is likely to become more—not less—strategically salient.

By 2025–2026, the strait's geophysical determinants create a narrow corridor—13–14 kilometers at its pinch point—with strong density-driven currents and variable depths that impose surface compression while still permitting deepwater submarine access. Localized ecological-operational risks, including recurring orca interactions with small craft, add an additional layer of unpredictability. Its strategic-military geometry is shaped by UK control of the Gibraltar base within a wider NATO environment, overlaid with Spanish contestation and frequent transits by external navies. The result is a layered deterrence architecture anchored in observation, access, and continuous presence.

Economic and bio-industrial flows combine dense through-traffic with regional transshipment and energy-modality constraints—ports, regasification capacity, and pipeline infrastructure—that limit near-term substitution. Cross-strait mobility and the convergence of cables and pipelines introduce hybrid constraints: when diversion occurs, the burden shifts into terminal capacity, infrastructure protection, and digital-system resilience. Governance rests on a stable but contested sovereignty equilibrium among the UK, Spain, and Morocco, now reframed by the 2026 Gibraltar treaty, which removes the land-border fence and shifts Schengen controls to Gibraltar's port and airport under Spanish responsibility.

Demographic and civilizational interfaces reinforce Gibraltar's role as a Europe–Africa boundary, with migration pressure shaping political narratives and security postures on both shores. Meanwhile, technological infrastructures—undersea cables, energy pipelines, AI-enabled surveillance systems, and smart-port architectures—create a high-frequency interface where gray-zone activity and cyber-nautical disruption can impose costs without requiring physical closure.

In this configuration, Gibraltar functions not merely as a passage but as a continuously observed, densely layered, hybrid chokepoint whose stability depends on the interplay of geography, governance, and technological resilience.

This chapter demonstrates how the SSA model reveals the

internal logic of a major strait whose significance is rooted in geography, history, and the enduring dynamics of civilizational interaction. The analysis now moves eastward to the Bosporus and Dardanelles, where the dynamics of empire, identity, and strategic access converge in one of the most symbolically and geopolitically charged chokepoint complexes in the world.

11

Chapter 11

The Taiwan Strait: The World's Most Dangerous Waterway

The Taiwan Strait is the most consequential maritime flashpoint in the global system because it fuses extreme military compression with extreme economic dependency. It is a corridor of routine trade and data flow, but also a frontier between political orders whose stability is central to Indo-Pacific deterrence. In the Straits Systems Architecture (SSA) framework, Taiwan's significance derives from the tight coupling of strategic-military geometry with technology supply chains, alliance commitments, and civilizational narratives. This chapter analyzes the strait as a system whose disruption would propagate far beyond East Asia, reshaping global manufacturing, finance, and security posture. The strait is not merely a body of water; it is a structural node whose compression of military, economic, informational, and identity systems makes even limited coercion potentially systemic.

Military Compression, Gray-Zone Erosion, and the Silicon Shield

The geophysical determinants of the Taiwan Strait create a narrow maritime corridor—roughly 126 kilometers at its narrowest points—whose shallow bathymetry, strong currents, and complex acoustic environment shape both commercial navigation and military operations. These physical features compress surface traffic while offering limited hiding places for submarines, turning geography into a multiplier for surveillance and interdiction. The strait's eastern approaches also feature deeper-water channels that facilitate submarine transit between the Pacific and the South China Sea, adding a three-dimensional layer to its strategic geometry.

This geophysical compression is most intensely experienced in the frontline micro-geographies of the outlying islands. Kinmen lies only a few kilometers from the Chinese mainland, while Matsu is similarly proximate. These islands, administered by Taiwan, function as literal forward positions where sovereignty is contested through daily interactions. Following a fatal capsizing incident involving a Chinese speedboat in February 2024, the China Coast Guard (CCG) escalated its patrol and enforcement posture around Kinmen. By 2025 and into early 2026, CCG vessels repeatedly entered Taiwan-designated restricted waters in patterns that normalize Beijing's jurisdictional claims. These operations represent a deliberate gray-zone strategy of incremental erosion: each incursion tests response thresholds, normalizes presence, and shifts the practical baseline of control without triggering open conflict.

The strait's strategic meaning is therefore not uniform; it is most volatile precisely where geophysical proximity meets active sovereignty contestation.

The strategic-military geometry of the Taiwan Strait is defined by sustained gray-zone pressure and reduced warning time for escalation. In 2025, PLA aircraft conducted 3,764 air incursions into Taiwan's de facto Air Defense Identification Zone (ADIZ)—a record level and a 22.4 percent increase from 2024—normalizing a tempo of coercion that strains readiness and seeks to desensitize threat perception. Large-scale exercises around Taiwan in 2024, including Joint Sword-2024A and Joint Sword-2024B, reinforced the operational logic of encirclement and joint pressure while expanding the political utility of "routine" high-tempo activity as a form of signaling.

A critical and often under-appreciated distinction in this geometry is the difference between a China Coast Guard (CCG)-led "quarantine" and a People's Liberation Army (PLA)-led "blockade." In a quarantine scenario, Beijing asserts domestic law-enforcement jurisdiction, requiring vessels bound for Taiwan to submit to inspections and administrative procedures. Framed as regulatory enforcement rather than an act of war, this tactic can paralyze shipping and insurance markets without crossing the political threshold of a formal military blockade. It weaponizes legal ambiguity within the governance subsystem to achieve systemic disruption through non-kinetic means. A full PLA blockade, by contrast, would involve overt military interdiction, mine-laying, and the potential use of force against non-compliant vessels—actions that would sharply raise escalation risks and increase the likelihood of external military response. The distinction matters because it gives Beijing a calibrated ladder of coercion that complicates decision-making for Taiwan and its partners.

This military compression is inseparable from Taiwan's role as the world's most advanced semiconductor foundry. Taiwan Semiconductor Manufacturing Company (TSMC) held roughly 70 percent of global foundry revenue share in 2025, and its dominance is even more pronounced at the most advanced nodes that power AI, high-performance computing, and modern defense systems. The "silicon shield" thesis holds that this economic centrality deters invasion because the costs of disruption would cascade globally. Yet the shield is double-edged: it also makes Taiwan a high-value target for coercion short of invasion—through quarantine, selective denial, or precision disruption of critical infrastructure. The Taiwan Strait therefore functions simultaneously as a commercial artery and as a potential trigger for great-power conflict whose economic consequences would dwarf those of any other contemporary chokepoint.

Economic Dependency, Technological Centrality, and Existential Vulnerabilities

The economic and bio-industrial flows through the Taiwan Strait are defined by extreme concentration and limited substitutability. Beyond routine container and bulk trade, the corridor sits adjacent to the dense supply chains of East Asian advanced manufacturing and carries critical components, equipment, and specialized materials that sustain global electronics production. Taiwan's dominance in leading-edge semiconductors means that even temporary disruption of production or export flows would trigger immediate shortages in AI servers, smartphones, automotive chips, and defense electronics worldwide. The dependency is asymmetric: advanced economies in the United

States, Europe, Japan, and South Korea have few immediate alternatives for the most sophisticated chips, while China remains heavily reliant on Taiwan's ecosystem for its own technology sector.

This concentration creates sharp modality constraints. Unlike many energy flows that can sometimes be rerouted via pipelines or alternative sea lanes, advanced semiconductor production cannot be quickly replicated or relocated. Fabrication facilities require years to build and qualify; specialized equipment and materials have concentrated supply chains; and the process know-how embedded in Taiwan's ecosystem is not readily transferable. Alternative routing offers little relief for time-sensitive, high-value components. In this sense, the Taiwan Strait is not merely a transit corridor but a production and innovation node whose disruption would impose non-linear shocks on global technology supply chains, inflation, and industrial output.

Taiwan's most acute import vulnerability lies in energy. The island imports nearly all of its natural gas and relies heavily on liquefied natural gas (LNG) to supply roughly half of its electricity generation. Taiwan's gas security-stockpile requirement has been set on an upward glidepath—11 days by 2025 and 14 days by 2027—but even a two-week buffer is thin under crisis conditions. In a prolonged coercion scenario, a sustained chokehold at sea would not merely stop chips from leaving Taiwan; it would threaten the continuous energy supply required to power fabrication plants and to sustain civilian infrastructure, hospitals, water systems, and transportation networks. Energy starvation could degrade semiconductor output and social endurance long before physical destruction of facilities occurs, undermining the "silicon shield" from within.

CHAPTER 11

Governance, Identity, and Hybrid Risks in a Contested Frontier

The governance regime of the Taiwan Strait is defined by contested sovereignty and the deliberate bracketing of ultimate political questions in favor of functional stability. A useful analogue comes from Taiwan's fisheries arrangement with Japan in disputed waters near the Senkaku/Diaoyutai Islands. By establishing a shared management area, joint regulatory mechanisms, and explicit sovereignty bracketing, the two sides reduced routine collision risk between coast guards and fishing fleets while preserving formal positions. The strategic logic is transferable: separating access, enforcement, and resource management from ultimate territorial claims can reduce day-to-day escalation pathways and produce a stabilizing "state effect" through routine administration and negotiation. For the Taiwan Strait system, this underscores that governance is not only about transit rights and military signaling; it is also about the legal-institutional engineering of contested maritime space through functional regimes that stabilize activity while leaving sovereignty disputes formally unresolved.

Civilizational and demographic interfaces add another layer of intensity. Longitudinal polling by National Chengchi University's Election Study Center shows that exclusive "Taiwanese only" identification has risen over three decades—from 17.6 percent in the early 1990s to roughly the low-60-percent range by the mid-2020s—with higher figures among younger cohorts. This shift reflects generational change, democratic experience, and reactions to Beijing's conduct, hardening the identity boundary across the strait. The civilizational dimension influences escalation thresholds: narratives of distinct

Taiwanese identity reinforce political resistance to unification under Beijing's terms, while Chinese narratives frame Taiwan as an inalienable part of the motherland whose separation is a historical humiliation. These identity dynamics interact with military geometry and economic stakes to make compromise more difficult and miscalculation more dangerous.

Technological infrastructures overlay additional hybrid risks. Taiwan's undersea cables have experienced repeated disruptions, including incidents linked to ship anchoring in nearshore waters and natural hazards such as earthquakes and submarine landslides offshore. Taiwan's Ministry of Digital Affairs has assessed that anchor dragging has been a leading cause of nearshore cable incidents in recent years, and it has shifted policy emphasis from passive repair toward proactive defense and resilience building. In parallel, Taiwan has accelerated investment in redundant communications pathways. Chunghwa Telecom secured Taiwan's first commercial license to provide Low Earth Orbit (LEO) satellite services using Eutelsat OneWeb, reflecting a deliberate strategy to add orbital resilience that can bypass compromised subsea infrastructure. Cyber-nautical threats, AI-enabled surveillance and electronic warfare, and potential seabed operations further complicate the picture: control over the informational environment of the strait is becoming as strategically significant as control over its physical waters.

The Taiwan Strait as of 2025–2026 presents a uniquely dense SSA configuration. Geophysical determinants impose narrow maritime compression with complex acoustic and current conditions that favor surveillance and interdiction, while also creating frontline micro-geographies (Kinmen and Matsu) where sovereignty is eroded through routine gray-zone incursions.

Strategic-military geometry features sustained high-tempo coercion (including 3,764 ADIZ incursions in 2025) alongside dense A2/AD capabilities, the credible shadow of great-power intervention, and the deliberate use of CCG-led quarantine tactics that weaponize legal ambiguity to impose systemic disruption without crossing the threshold of overt blockade. Economic and bio-industrial flows center on irreplaceable semiconductor production capacity that creates extreme global dependency and sharp modality constraints, while Taiwan's limited LNG buffer represents an existential energy vulnerability that could degrade fabrication capacity and civilian endurance before physical destruction occurs. Governance regimes rely on functional, sovereignty-bracketing arrangements that stabilize routine activity while leaving ultimate political questions unresolved. Demographic and civilizational interfaces sustain a hardening identity boundary that interacts with military and economic stakes to raise escalation risks. Technological infrastructures add layered hybrid vulnerabilities through undersea cables, cyber exposure, and the accelerating deployment of LEO satellite redundancy as a countermeasure. The result is a system in which military, economic, informational, identity, and governance pressures are tightly coupled, making even calibrated coercion potentially systemic in its consequences.

This chapter demonstrates how the SSA model reveals the internal logic of the world's most dangerous waterway—a strait whose significance is rooted in the fusion of military compression, technological centrality, existential energy vulnerability, and unresolved political identity. The analysis now moves westward to the Bosporus and Dardanelles, where the dynamics of empire, identity, and strategic access converge in one of the most symbolically and geopolitically charged chokepoint complexes in the world.

12

Chapter 12

The Bosporus & Dardanelles: The Ottoman Legacy

Geophysical Compression and Historical Gateways

The Bosporus and Dardanelles form a dual-strait system that connects the Black Sea to the Mediterranean and, by extension, to the global maritime domain. Together, they constitute one of the most symbolically charged and strategically consequential chokepoint complexes in the world. Their significance derives not only from geography but from history: for centuries, control of these straits defined the structure of regional order, shaped the rise and decline of empires, and conditioned the strategic imagination of states across Eurasia. The legacy of the Ottoman Empire continues to shape the governance, identity, and strategic behavior surrounding these narrow corridors. In 2025, 40,172 commercial vessels transited the Bosporus, and persistent Black Sea incidents—including drone attacks on oil tankers near the strait's approaches in 2026—underscored the system's enduring volatility.

The geophysical determinants of the Bosporus and Dardanelles establish the constraints under which all other systems operate. The Bosporus is exceptionally narrow—at its tightest point only about 700 meters wide—with sharp turns, strong currents, and complex hydrological dynamics driven by the salinity and density differences between the Black Sea and the Mediterranean. The Dardanelles is wider but still constrained, with a long, linear geometry that channels maritime movement. Both straits are deep enough to accommodate large vessels but narrow enough to create predictable trajectories and limited maneuverability. Their physical characteristics compress maritime movement into confined corridors that are inherently difficult to secure and easy to monitor. These constraints also concentrate ecological externalities: the high-traffic, high-current environment elevates the risk of collisions, groundings, and spills that could severely impact Black Sea and Marmara fisheries as well as the sensitive ecosystems of the straits themselves. Geography does not merely shape access; it structures the strategic logic of the entire Black Sea region.

States have long sought to engineer their way around these constraints. Turkey's Kanal Istanbul project envisions a 45-kilometer artificial waterway bisecting the European side of Istanbul to connect the Black Sea directly to the Sea of Marmara. While actual canal excavation has faced repeated delays into 2025 amid fierce environmental opposition and financing constraints, large-scale urban development and real estate activity in the surrounding "Yenişehir" zone have advanced. Turkish officials have framed the project as a sovereign bypass that could, in theory, operate outside the strictures of the Montreux

Convention. This effort illustrates a recurring pattern in choke-point politics: when physical geography imposes unacceptable constraints, states attempt to construct new geography to expand optionality, even as legal, environmental, and financial costs accumulate.

The deeper history of these waterways reinforces their role as gateways whose opening or closure can reorganize entire systems. In deep time, the connection between the Black Sea and the Mediterranean has been intermittent, shaped by glacial cycles, sea-level changes, and tectonic activity. The Bosporus and Dardanelles as we know them today are relatively recent geological features whose stability has conditioned the economic and ecological character of the Black Sea basin for millennia.

Economic Flows, Montreux Leverage, and Strategic Asymmetry

The economic and bio-industrial flows that pass through the Bosporus and Dardanelles are central to their global significance. The straits carry grain exports from Ukraine and Russia and remain a structural hinge for Black Sea access to global markets. They also channel energy shipments from the wider Caspian and Black Sea system and manufactured goods moving between Europe and Eurasia. These flows are critical for global food security, and disruption—whether from conflict, accidents, or deliberate interdiction—can create immediate supply shocks, particularly in food systems. Even after the collapse of the Black Sea Grain Initiative, significant volumes continued to move through alternative arrangements, with Turkey's facilitation and the operational realities of the straits remaining central to what is feasible at scale.

This economic centrality is inseparable from Turkey's strategic leverage under the Montreux Convention. The convention grants Ankara control over the transit of naval vessels, imposes tonnage and duration limits on non-Black Sea powers, and establishes distinct rules for wartime and peacetime passage. In wartime or when Turkey perceives a threat, it retains the authority to restrict belligerent warships—a right Ankara has exercised consistently since Russia's full-scale invasion of Ukraine in 2022. Recent incidents, including drone attacks on commercial tankers in the Black Sea approaches in 2026, heightened operational risk without altering Turkey's baseline enforcement posture. The straits are not merely passages; they are instruments of strategic regulation that allow Turkey to shape the military balance in the Black Sea while preserving merchant traffic.

A particularly consequential dimension of this regime emerged during the war in Ukraine. Under Article 12 of the Montreux Convention, Black Sea states may move submarines through the straits for the purpose of rejoining their base for the first time after construction or purchase, or for repair in dockyards outside the Black Sea. Prior to 2022, Russia exploited this allowance to rotate submarines between the Black Sea and the Mediterranean. When Turkey closed the straits to belligerent warships in 2022, Russia's Black Sea Fleet became far more constrained. The inability to reinforce or rotate certain assets at will became a long-run degradation mechanism—an illustration of "Montreux leverage" in which a near-century-old legal framework shapes the operational sustainability of a modern naval campaign.

CHAPTER 12

Governance, Civilizational Legacy, and Hybrid Risks

The governance regime of the Bosporus and Dardanelles is defined by the Montreux Convention, a legal framework that has endured for nearly a century. The convention reflects the legacy of the Ottoman Empire and the geopolitical realities of the interwar period, granting Turkey formal authority over warship transit while guaranteeing freedom of passage for merchant vessels. This governance structure shapes contemporary strategic dynamics. Turkey's authority gives it bargaining power in its relations with Russia, NATO, and the European Union, allowing Ankara to calibrate access, signal neutrality, and manage escalation in the Black Sea without relinquishing sovereign control over the straits. Governance is therefore both legal and geopolitical: the straits operate under a formal regime that is interpreted through the lens of contemporary power politics and crisis management.

The demographic and civilizational interfaces surrounding the straits add depth to their strategic significance. Istanbul, situated on both sides of the Bosporus, is a city shaped by centuries of migration, empire, and cultural synthesis. The straits mark the boundary between Europe and Asia, between Christian and Islamic civilizations, and between the legacies of Byzantium and the Ottoman Empire. These civilizational dynamics influence political identities, historical narratives, and the symbolic meaning of the straits. Control of the Bosporus and Dardanelles is not only a strategic asset; it is a civilizational marker embedded in Turkey's national identity and historical memory. This symbolic dimension shapes the willingness of states to assert control, negotiate compromise, or escalate in defense of perceived core interests.

The technological infrastructures that support maritime movement through the straits are increasingly central to their strategic dynamics. Undersea cables—including KAFOS and the Caucasus Cable System—connect Turkey with Black Sea and Caucasus nodes, while wider regional systems such as MedNautilus link onward into the Mediterranean. These infrastructures, combined with dense surveillance, vessel traffic management, and pilotage systems, are essential for navigating narrow, high-current waterways in a megacity environment. Cyber-nautical risks and the possibility of sabotage—whether accidental or deliberate—introduce new forms of vulnerability in high-traffic corridors where even short interruptions can cascade into queueing and economic cost. This exposure is compounded by the growing relevance of seabed operations and port digitization: Unmanned Underwater Vehicles (UUVs) and other seabed capabilities create new gray-zone tools for monitoring and disruption, while terminals and cable landing stations face exposure through smart-port operational technology (OT) systems that control automated equipment and logistics flows. The straits are not only physical corridors but digital ones, and control over their informational environment is becoming a form of strategic leverage.

Hybrid Pressures and the Limits of Legal Leverage

The governance and economic subsystems of the straits have been tested by the hybrid dynamics of the war in Ukraine. Since 2022, floating naval mines have repeatedly broken loose from their moorings during storms and drifted south toward the Bosporus, creating acute operational anxiety and collision risk in a channel that is already among the world's most demanding for navigation. In response, NATO members Turkey, Romania, and Bulgaria activated the Mine Countermeasures (MCM) Black Sea Task Group in July 2024. This ad hoc, trilateral maritime policing coalition illustrates how littoral states are compelled to forge practical cooperation to protect trade and bio-industrial flows against hybrid hazards that no single actor can fully contain.

At the same time, Turkey has demonstrated how administrative and insurance requirements can become instruments of chokepoint management. In December 2022, amid the rollout of sanctions and price-cap regimes on Russian seaborne oil, Ankara demanded explicit Protection and Indemnity (P&I) insurance letters from tankers transiting the Turkish Straits to reduce the catastrophic liability exposure of a spill in a narrow urban waterway. The policy triggered tanker queues and delays at the straits' approaches. The episode illustrated a modern form of "paper blockade": regulatory and insurance demands can disrupt flows and reprice risk without a single warship firing a shot, while also functioning as a defensive measure to protect the host state from outsized environmental and political consequences.

The Bosporus and Dardanelles form a Straits Systems Architecture defined by exclusivity, historical depth, and legal-strategic complexity. Their significance arises from the fact that they constitute the **sole maritime access point to the Black Sea**, a position that embeds them in the legacies of empire and places them under a governance regime unlike any other chokepoint in the world. Their vulnerabilities emerge from geopolitical competition, civilizational narratives, and the accelerating integration of digital infrastructures that now shape maritime governance as much as physical geography.

By 2025–2026, the geophysical determinants of the Turkish Straits create one of the most compressed navigational environments on earth. The Bosporus narrows to roughly **700 meters** at its tightest point, with sharp turns, strong bidirectional currents, and complex hydrology that force vessels into predictable but hazardous trajectories. The Dardanelles, by contrast, is a long, linear channel whose length amplifies the consequences of any disruption. Together, these waterways concentrate ecological and operational risks, prompting Turkey to pursue ambitious engineered bypasses such as **Kanal Istanbul**—a project intended to relieve pressure but which itself introduces new geopolitical and environmental debates.

The strategic-military geometry of the straits is dominated by the **Montreux Convention of 1936**, which grants Turkey structural gatekeeping power over naval access between the Black Sea and the Mediterranean. Montreux imposes strict tonnage, duration, and notification requirements on non-littoral warships, creating an asymmetry that limits external naval power while shaping escalation dynamics with Russia, NATO, and the European Union. This legal architecture becomes especially consequential during crises: Turkey's invocation of Montreux provisions after the outbreak of the Russia–Ukraine war restricted the movement of Russian naval assets and constrained Moscow's ability to rotate submarines and reinforce its Black Sea fleet. At the same time, the drifting mines and hybrid threats that proliferated in the Black Sea forced the creation of ad hoc coalitions—such as the **MCM Black Sea Task Group**—to manage hazards that Montreux was never designed to address.

Economic and bio-industrial flows through the straits exceed **40,000 vessels per year**, including critical grain and energy exports that make the Bosporus–Dardanelles complex a global food and energy security node. Disruption here produces rapid commodity shocks, as demonstrated during the interruptions of Black Sea grain flows in 2022–2024. Even administrative measures—such as insurance requirements, pilotage rules, or environmental liability protocols—can be weaponized to slow traffic, impose costs, or create de facto disruptions without formally closing the straits. The system's sensitivity is heightened by the fact that the straits are not merely transit corridors but the connective tissue between inland production zones and global markets.

Governance is anchored in the institutionalized leverage of the Montreux Convention, which provides Turkey with both responsibility and bargaining power. Montreux's flexibility under crisis conditions allows Ankara to calibrate access in ways that reflect its strategic priorities, while its submarine provisions have repeatedly constrained Russia's ability to sustain forward deployments. This governance regime is not static; it is continually interpreted through the lens of Turkey's national identity, regional ambitions, and its role as a balancing power between NATO and Eurasian actors.

Demographic and civilizational interfaces further deepen the straits' complexity. Istanbul sits at the meeting point of Europe and Asia, Christianity and Islam, empire and republic. It is an imperial synthesis and a national symbol, and its identity shapes strategic commitments and escalation thresholds. The straits are not only physical boundaries but civilizational markers, where narratives of sovereignty, historical memory, and cultural continuity influence governance choices and diplomatic behavior.

Technological infrastructures add yet another layer of hybrid exposure. Undersea cables—including systems such as

KAFOS—converge in the region, linking Europe, the Caucasus, and the Middle East. Dense surveillance networks, vessel-traffic management systems, and AI-enabled monitoring tools create a digital overlay that is now inseparable from physical navigation. These systems introduce new disruption modes: cyber-nautical interference, data manipulation, and gray-zone activity that can impose costs or create uncertainty without any overt military action. In channels as compressed as the Bosporus and Dardanelles, even minor digital disruptions can cascade into significant operational consequences.

Taken together, the Bosporus and Dardanelles function as a uniquely exclusive, historically layered, legally codified, and technologically hybrid chokepoint system. Their stability depends on the interplay of geography, law, identity, and digital infrastructure—an interplay that continues to shape the strategic balance of the Black Sea, the Mediterranean, and the wider international system.

The straits are not merely narrow passages; they are structural hinges in the Eurasian system whose stability influences the strategic behavior of regional and global powers. This chapter demonstrates how the SSA model reveals the internal logic of a dual-strait system whose significance is rooted in history, identity, and the geopolitics of access. The analysis now moves northward to the Bering Strait, where the dynamics of climate change, Arctic emergence, and shifting geopolitical boundaries converge in a new and rapidly evolving chokepoint.

13

Chapter 13

The Bering Strait: The Arctic Emerges

The Bering Strait is the narrowest point between the United States and Russia and the southern gateway to the Arctic Ocean. For most of modern history, it occupied a position of strategic marginality—a remote corridor defined by ice, distance, and limited economic activity. That marginality is now ending. Climate change, shifting energy patterns, emerging Arctic shipping routes, and renewed great-power competition are transforming the Bering Strait into a chokepoint of growing global significance. It is becoming a new strategic frontier, where geography, technology, and geopolitics converge in a rapidly evolving configuration.

By 2025, the Northern Sea Route (NSR) carried **37.02 million metric tons of cargo**, an all-time high, while vessel transits through the Bering Strait continued to rise sharply as the navigable season lengthened. Transit counts vary by definition and AIS coverage, but Marine Exchange of Alaska data shows an increase from **242 recorded transits in 2010** to **665 in 2024**, illustrating how Arctic access is converting environmental change into measurable traffic growth. Record-low sea ice in late 2025 and early 2026 further extended the navigable window, reinforcing the structural shift underway.

The geophysical determinants of the Bering Strait define both the constraints and opportunities of its emerging role. The strait is narrow—about **85 kilometers** at its tightest point—and shallow, averaging **30–50 meters** in depth. Historically, these dimensions made the corridor seasonally hostile: severe storms, drift ice, and limited charting and response infrastructure suppressed reliable navigation. Climate change is now altering that baseline. In 2025, Arctic sea ice reached a low summer minimum consistent with the post-2007 "new Arctic" regime, while the winter maximum remained historically weak. By early 2026, Arctic winter sea-ice extent set a new record low. The result is structural rather than episodic: the navigable season is lengthening from a short summer window toward multiple months, and the Bering Strait is becoming a more reliable gateway—not because the strait itself has changed, but because the environment around it has.

These geophysical shifts carry significant ecological externalities. Reduced sea ice and longer navigable seasons increase the risk of ship strikes on marine mammals, elevate underwater noise that disrupts subsistence hunting, and heighten the danger of fuel spills in a region with limited response capacity. As shipping volumes grow, the cumulative environmental footprint of the corridor rises even as the physical geography remains relatively stable. The Bering Strait's transformation is therefore not purely a story of opportunity; it is also one of heightened ecological vulnerability in a fragile and rapidly changing marine environment.

The deeper history of Arctic gateways reinforces a broader structural principle: changes in accessibility reorganize entire systems. In deep time, the Bering Strait alternately connected and separated continents and ecosystems, shaping patterns of migration, climate, and biological exchange. Its current opening is driven by human-induced warming rather than natural cycles, yet the underlying lesson remains the same. When a narrow passage becomes reliably navigable, it alters trade routes, resource access, military posture, and the balance of power across entire regions. The Bering Strait is now entering such a phase—its strategic meaning expanding as environmental thresholds shift and global competition intensifies.

Military Geometry and Strategic Proximity

The strategic-military geometry of the Bering Strait reflects the renewed centrality of the Arctic in great-power competition. The strait sits between Alaska and Russia's Chukotka region, placing it at the intersection of two major military theaters and compressing warning time by geography alone. Russia has expanded Arctic infrastructure and patrol activity and treats the Northern Sea Route as a strategic corridor. China, increasingly aligned with Russia, has begun to operate in the Pacific Arctic as well. In July 2024, Russia and China conducted a joint strategic bomber patrol over the Chukchi and Bering Seas, involving Russian Tu-95MS and Chinese H-6 bombers that were intercepted by U.S. and Canadian fighters in the Alaska ADIZ. In late 2024, two China Coast Guard cutters—escorted by Russian border-guard vessels—were observed operating in the Bering Sea in the northernmost area where U.S. observers had previously seen Chinese coast-guard ships. These events mark a threshold shift: the Bering Strait is no longer treated as a marginal seam, but as a theater in which major powers signal presence and test surveillance and response architectures.

Economic Emergence, Governance Strain, Indigenous Interfaces, and Hybrid Risks

The economic and bio-industrial flows emerging around the Bering Strait are central to its rising strategic significance. As Arctic sea ice recedes, the Northern Sea Route (NSR) and, eventually, trans-polar routing offer shorter shipping distances between Asia and Europe, reducing voyage times under favorable conditions compared with traditional Suez or Panama passages. Yet the NSR's core trade today is not international transit but Russian Arctic export logistics: total cargo shipments reached **37.02 million metric tons in 2025**, driven primarily by LNG, oil, and condensate. The Bering Strait is the mandatory gateway for vessels moving between the Pacific and the Arctic, including NSR traffic and Northwest Passage transits. Vessel counts vary by definition and AIS coverage, but recorded transits have risen sharply throughout the 2020s, reflecting expanding seasonal activity and the growing strategic salience of the corridor.

This economic emergence carries dangerous and illicit dimensions. Western sanctions targeting Russia's Arctic LNG expansion have restricted access to specialized icebreaking LNG carriers and associated logistics, pushing Moscow toward irregular shipping practices. A 2025 analysis by the Bellona Environmental Transparency Center found that **100 sanctioned**

vessels sailed along Russia's Arctic coast—nearly a third of the cargo ships operating the route—many of them older tankers with weak insurance and limited ice-class capability. The Bering Strait thus becomes a funnel not only for legitimate seasonal commerce but also for heightened spill and collision risk in one of the world's most environmentally fragile operating environments.

Governance in the Bering Strait is shaped by national sovereignty, international law, and evolving Arctic institutions. The United States and Russia share responsibility for the corridor under UNCLOS-based navigation norms, but cooperation has been strained by the geopolitical rupture following Russia's invasion of Ukraine. This has weakened the operational effectiveness of multilateral institutions such as the Arctic Council and left governance in a transitional state—anchored in formal legal frameworks yet increasingly challenged by new environmental, economic, and military realities. In response to deepening Sino-Russian activity in the region, the U.S. Department of Defense issued its 2024 Arctic Strategy and adopted a "monitor-and-respond" approach built on urgent investments in all-domain awareness and improved early warning. Yet the U.S. response faces a material constraint: Alaska has historically lacked a true Arctic deepwater port capable of supporting sustained Coast Guard and naval presence. The Port of Nome deep-draft expansion—now moving into construction planning—signals an attempt to close that infrastructure gap, but until such capacity becomes operational, geography continues to outrun logistics.

The demographic and civilizational interfaces surrounding the Bering Strait add further depth to its strategic meaning. Indigenous communities—including the Yupik, Inupiat, and

Chukchi—have inhabited the region for millennia, maintaining cultural, economic, and familial ties across the strait. These communities are directly affected by environmental change, resource development, and geopolitical tension. Their presence introduces a human dimension to the strait's strategic environment, shaping governance legitimacy, environmental stewardship, and the social foundations of Arctic security. Climate-driven changes in sea ice, marine mammal migrations, and weather patterns affect subsistence practices, while increased shipping and military activity raise concerns about consultation, sovereignty, and the protection of Indigenous rights in a rapidly transforming region.

Technological infrastructures supporting Arctic navigation and surveillance are becoming increasingly central to the Bering Strait's strategic dynamics. Satellite monitoring, autonomous sensors, ice-forecasting algorithms, and resilient communications are essential for safe transit in a region where weather, darkness, and remoteness magnify risk. As economic activity grows, proposals for expanded undersea cables and energy infrastructure will introduce new dependencies and new hybrid vulnerabilities. Cyber-nautical risks and data-integrity challenges matter more in the Arctic than in most corridors because redundancy is scarce and response times are long. To mitigate the risks of remoteness and potential infrastructure disruption, Arctic actors are increasingly turning to Low Earth Orbit (LEO) satellite constellations for resilient, low-latency communications and positioning that can supplement—or bypass—vulnerable terrestrial systems. Control over the informational environment of the strait will therefore be a key determinant of future chokepoint stability.

The Bering Strait exemplifies a Straits Systems Architecture defined by **emergence**. Its significance is not inherited from history but produced by transformation—environmental, technological, and geopolitical. Its vulnerabilities are shaped by climate change, great-power proximity, and the fragility of Arctic governance. By 2025–2026, the strait's geophysical determinants create a narrow and shallow corridor whose weakening sea-ice constraints are transforming a historical barrier into a seasonal gateway while simultaneously concentrating ecological risk. Its strategic-military geometry reflects U.S.–Russia proximity across the strait, Russia's expanding Arctic posture, and deepening Sino-Russian cooperation—including joint bomber and coast-guard operations near Alaska—that introduce triangular competition and heightened surveillance demands. Its economic and bio-industrial flows reflect a nascent but accelerating Arctic routing system, with the NSR carrying roughly 37 million metric tons of cargo in 2025 and the Bering Strait serving as the mandatory Pacific–Arctic funnel, even as sanctions pressure pushes a growing "shadow fleet" of poorly insured vessels into Arctic waters. Its governance regime rests on UNCLOS-based frameworks and Arctic cooperation mechanisms under strain, while the U.S. "monitor-and-respond" doctrine faces a modality constraint in Alaska's limited deepwater port and replenishment capacity. Its demographic and civilizational interfaces center Indigenous cross-strait communities whose subsistence and cultural continuity are directly affected by shipping, noise, and ecological change. And its technological infrastructures integrate satellites, autonomy, ice forecasting, and emerging connectivity plans, generating hybrid vulnerabilities intensified by remoteness and data dependence—partly mitigated by the growing role of LEO satellite redundancy.

14

Chapter 14

The Danish Straits: Europe's Northern Gate

The Danish Straits—the Øresund, the Great Belt, and the Little Belt—form the maritime threshold between the Baltic Sea and the North Sea. They are the only navigable access points for the states of the Baltic region, making them a structural hinge in the security architecture of Northern Europe. Unlike the volatility of the Bab el-Mandeb or the militarized tension of the Taiwan Strait, the Danish Straits are defined by stability, institutiona lization, and the dense integration of European political and economic systems. Yet their stability does not diminish their strategic significance. In 2025 the three straits collectively handled approximately 75,000 vessel transits, with the Great Belt alone recording roughly 27,000 commercial ships and over 8,000 full transits (Skagen–Bornholm). Hydrocarbon traffic has increased markedly since 2022, reflecting the region's role as a critical energy corridor amid Europe's transition away from Russian pipeline gas. The straits are a chokepoint whose importance is increasing as the Baltic becomes a focal point of energy transition, military realignment, and great-power competition.

Geophysical Constraints and Regional Connectivity

The geophysical determinants of the Danish Straits establish the constraints under which all other systems operate. The straits are narrow, shallow in places, and characterized by complex currents and variable salinity. The Great Belt offers the deepest route (often exceeding 20 meters in the main fairway) but still imposes draft and maneuverability limits; the Øresund is narrower and shallower. Ice conditions, once a significant constraint, have diminished in recent decades, increasing the navigable season and altering winter-risk profiles. The physical geometry compresses maritime movement into predictable corridors, creating a natural regulatory mechanism for access to the Baltic. These constraints also concentrate ecological externalities: higher traffic density and growing hydrocarbon and LNG movements elevate spill and emissions risk in a semi-enclosed sea whose ecosystems are already under pressure from nutrient loading and climate-driven change. The deeper role of these waterways as connectors rather than barriers is reinforced by their integration into broader European networks, including the Øresund Bridge, which physically embodies the permeability of the region while the straits serve as its maritime counterpart.

The strategic-military geometry of the Danish Straits reflects the evolving security environment of Northern Europe. The straits are the only maritime access point for the navies of Russia, Finland, Sweden, Poland, Germany, and the Baltic states. Denmark's control over the straits, combined with its NATO membership, gives the alliance a structural advantage in regulating access to the Baltic Sea. Finland's accession (2023) and Sweden's accession (2024) have transformed the Baltic into a predominantly NATO-aligned maritime space—often described as a "NATO lake." This consolidation has enabled enhanced operations to deter and monitor hybrid activity. In January 2025, NATO launched Baltic Sentry, and Allied Command Transformation's Task Force X-Baltic has demonstrated how uncrewed systems and AI-enabled surveillance can expand persistent monitoring of critical undersea infrastructure. The straits therefore function as a strategic valve: they can enable or constrain naval movement, influence deterrence dynamics, and shape the operational environment of the region.

With Finland and Sweden in NATO, Russia's Baltic Fleet assets in Kaliningrad and the Gulf of Finland operate under conditions of near-encirclement. The Danish Straits are therefore not only a corridor for Russian oil exports; they are Russia's key maritime lifeline for resupplying the Kaliningrad exclave should overland movement through Poland and Lithuania be constrained. This geography helps explain the attraction of low-cost, deniable hybrid tactics in the wider Baltic—GNSS interference, intimidation signaling, and sabotage of undersea infrastructure—which can impose political and economic costs without requiring sustained conventional sea control.

Economic and Energy Flows in the European Transition

The economic and bio-industrial flows that pass through the Danish Straits are central to their global significance. The straits carry container traffic, energy shipments (including LNG imports to support Europe's post-Russian-gas transition), and bulk commodities between the Baltic states and global markets. They are also conduits for the export of grain, timber, and industrial goods from the region. The energy transition has elevated their importance as routes for offshore wind components, emerging hydrogen networks, and the maritime logistics of Europe's green shift—including joint Denmark–Germany projects such as the Bornholm Energy Island (3 GW offshore wind hub agreed in early 2026). The dependency structure is asymmetric: Baltic states rely heavily on the straits for access to global markets, while global actors depend on the straits for specific commodities and industrial inputs. The straits are not merely passages; they are structural components of the economic architecture of Northern Europe.

These flows are increasingly shaped by the imperatives of decarbonization. The straits are becoming arteries for the components and materials required for offshore wind farms, grid interconnectors, and hydrogen infrastructure that underpin Northern Europe's electrification. In January 2026, Denmark and Germany concluded an agreement on the Bornholm Energy Island project—an offshore hub designed to connect 3 GW of offshore wind generation to both national grids—illustrating how the Baltic's physical logistics are being fused to Europe's energy-security and climate objectives. Disruption in the Danish Straits would therefore not only affect traditional trade but could slow the physical rollout of Europe's green transition, creating a new category of strategic vulnerability tied to climate goals rather than fossil-fuel dependence.

A particularly acute ecological and governance challenge has emerged from the composition of hydrocarbon traffic itself. New data compiled by the Danish Maritime Authority indicates that EU-sanctioned tankers linked to Russia's "shadow fleet" made 292 voyages through Danish waters in 2025. These aging vessels, often operating under opaque ownership and uncertain insurance, represent an outsized environmental threat in narrow, shallow channels. Denmark and EU partners have explored legal avenues to monitor, inspect, or constrain high-risk vessels for environmental and insurance non-compliance without violating the long-standing free-passage regime rooted in the Copenhagen Convention of 1857 and reflected in UNCLOS Article 35(c). The result is a modern chokepoint dilemma: how to preserve predictable passage while preventing the straits from becoming an unpoliced corridor for sanctions evasion and ecological catastrophe.

Governance, Military Geometry, and Hybrid Risks

The governance regime of the Danish Straits is among the most institutionalized in the world. The Copenhagen Convention of 1857 abolished tolls and established free commercial navigation through the Sound and the Belts, creating a legal foundation that predates most modern maritime regimes. Under UNCLOS, the Danish Straits are treated as a special category under Article 35(c), in which passage is regulated not by generic transit-passage rules but by long-standing conventions specific to these waterways. Denmark retains regulatory authority, yet exercises it within a dense framework shaped by European Union law, NATO commitments, and international maritime norms. Operational governance is reinforced by mandatory reporting and vessel-traffic services, including the IMO-recognized BELTREP ship-reporting system and the Great Belt VTS, with corresponding arrangements in the Øresund. This institutional stack converts congestion into managed flow and provides a platform for safety enforcement, environmental risk management, and hybrid-threat monitoring. Governance in the Danish Straits is not improvised; it is embedded in enduring legal commitments and reinforced by contemporary European institutions.

The demographic and civilizational interfaces surrounding the Danish Straits add depth to their strategic significance. The region sits at the intersection of Scandinavian, Germanic, and Baltic cultural spheres. Cities such as Copenhagen, Malmö, and Helsingør reflect centuries of maritime exchange, political rivalry, and cultural synthesis. The Øresund Bridge, linking Denmark and Sweden, symbolizes the region's integration and the permeability of its civilizational boundaries. These dynamics influence political identities, economic networks, and the social foundations of maritime security. The straits are not only strategic corridors; they are cultural interfaces shaped by cooperation, mobility, and shared institutions.

Technological infrastructures are increasingly central to the straits' strategic dynamics. Advanced traffic-management systems, digital logistics networks, and maritime-domain-a wareness platforms enable efficient and secure transit through narrow and heavily trafficked channels. Undersea cables, energy pipelines, offshore wind infrastructure, and emerging hydrogen corridors converge in the region, creating new dependencies and vulnerabilities. Recent hybrid disruptions have demonstrated how low-tech methods can generate high-impact effects. In November 2024, the Chinese-flagged bulk carrier *Yi Peng 3* came under investigation after two telecommunications cables—the BCS East-West Interlink and C-Lion1—were severed in the Baltic; Danish naval and maritime authorities monitored the vessel as it exited and anchored in the Kattegat. In December 2024, the tanker *Eagle S*, suspected of operating as part of Russia's shadow fleet, was linked to damage to the Estlink 2 power cable and multiple telecom cables in the Gulf of Finland via anchor dragging. These cases illustrate how sabotage and "accident" can blur in a dense seabed environment, and why Denmark functions as a physical gatekeeper: the straits provide not only access to the Baltic but also one of the few practical chokepoints where suspect vessels can be tracked, monitored, and, in coordination with partners, interdicted as they attempt to leave the theater.

The Danish Straits exemplify a Straits Systems Architecture defined by stability, institutionalization, and strategic asymmetry. Their significance derives from their role as the sole maritime access point to the Baltic, their embeddedness in European governance structures, and their centrality to the region's economic and energy systems. Their vulnerabilities are shaped by great-power competition, digital infrastructure, and the evolving dynamics of European security and energy transition. By 2025–2026, the geophysical determinants of the straits create narrow channels with variable depths and reduced ice constraints, producing predictable corridors that serve as natural regulatory mechanisms for Baltic access while concentrating ecological risks from rising hydrocarbon, LNG, and shadow-fleet traffic. The strategic-military geometry reflects a predominantly NATO-aligned Baltic following the accession of Finland and Sweden, enabling enhanced operations such as Baltic Sentry and Task Force X-Baltic while increasing Russia's incentive to rely on hybrid tactics under conditions of maritime encirclement. Economic and bio-industrial flows encompass roughly **75,000 annual transits**, including LNG and hydrocarbon shipments critical to Europe's post-Russian-gas transition, alongside offshore wind components and emerging hydrogen corridors that position the straits as an energy-transition artery. Governance remains anchored in the Copenhagen Convention of 1857, reinforced by EU and NATO frameworks and mandatory reporting systems such as BELTREP, producing institutionalized stability now being stress-tested by attempts to impose environmental and insurance compliance on shadow-fleet vessels without violating free-passage commitments. Demographic and civilizational interfaces sustain a Scandinavian–Germanic–Baltic synthesis centered on the Øresund integration zone. And technological infrastructures integrate advanced traffic management and maritime-domain awareness with dense undersea cable and energy networks, creating hybrid physical-digital-energy exposure made visible by recent anchor-dragging incidents and Denmark's role in monitoring vessels transiting out of the Baltic.

This chapter concludes the European flank of the applied case studies. The final case chapter turns to Panama, the freshwater engineered chokepoint whose climate sensitivity has reshaped interoceanic optionality. The book then moves to Part III, where the global architecture of chokepoints is synthesized into a unified model of systemic risk, strategic leverage, and future transformation.

15

Chapter 15

Panama: The Engineered Chokepoint

The Panama Canal stands as the most consequential engineered chokepoint in the global system. Unlike the geological corridors of Hormuz or Malacca, it is a human artifact whose geometry was imposed on the landscape rather than inherited from it. Its significance derives not from natural compression but from the decision to cut a narrow, lock-dependent passage between two oceans, thereby reshaping the economic geography of the Americas and the strategic geometry of global trade. In the decades since its opening, the canal has functioned as both an accelerator of commerce and a concentrated point of systemic vulnerability. The droughts of 2023–2025, which sharply reduced daily transits and forced widespread rerouting around the Cape of Good Hope, demonstrated that engineered straits can generate disruption on a scale comparable to their natural counterparts when environmental conditions turn against them.

CHAPTER 15

The Engineered Gateway and Its Climate Constraint

The geophysical determinants of the Panama Canal are defined by its artificial character. The canal relies on a freshwater system centered on Gatun Lake, which supplies the water required to operate its locks. This dependency makes the waterway acutely sensitive to rainfall variability. Prolonged drought reduces lake levels, forcing the Panama Canal Authority to impose draft restrictions and limit the number of daily transits. During the 2023–2025 crisis, transits fell dramatically below historical averages, creating backlogs that rippled through supply chains dependent on the Americas' east–west connection. The canal's expansion, completed in 2016 with larger locks and water-saving basins, increased capacity but did not eliminate its fundamental vulnerability to hydrological stress. Climate change is expected to intensify rainfall variability in the region, suggesting that the canal's reliability as a global artery will remain contingent on environmental conditions rather than solely on engineering or management.

This engineered character distinguishes Panama from natural straits. Where Hormuz or the Bosporus impose constraint through the given configuration of land and water, Panama imposes constraint through a system that must be continuously maintained and supplied with fresh water. The canal therefore exemplifies how human ambition can create new points of concentration whose risks are partly self-generated. When those risks materialize—as they did during the recent drought—the effects are not confined to Panama. They propagate through global commodity markets, particularly for grain, energy, and containerized goods moving between the Atlantic and Pacific basins.

Domestic ecological governance has become inseparable from the canal's operational stability. In late 2023, nationwide protests over the environmental impact of the Cobre Panamá copper mine led to its forced closure and the physical blockade of major highways. These actions paralyzed domestic logistics and exposed how watershed protection conflicts can directly constrain the arteries surrounding the canal. In an engineered chokepoint, civil society mobilization over resource extraction can freeze movement as effectively as drought or mechanical failure.

Economic Centrality and Strategic Competition

The economic and bio-industrial flows through the Panama Canal remain central to its global significance. The canal serves as a critical artery for container traffic, energy shipments, and agricultural commodities, particularly grain exports from the Americas to Asia and energy flows in both directions. Its disruption forces vessels into longer, more expensive routes around the Cape of Good Hope or, in some cases, through alternative land-bridge arrangements. The 2023–2025 restrictions demonstrated the canal's continued centrality: even with expanded capacity, the system lacked sufficient slack to absorb prolonged hydrological stress without generating widespread delay and cost increases.

States are now actively constructing physical alternatives to escape these modality constraints. Mexico has heavily invested in the Interoceanic Corridor of the Isthmus of Tehuantepec (CIIT), a rail land-bridge connecting the Pacific port of Salina Cruz to the Gulf port of Coatzacoalcos. By operationalizing this corridor in the mid-2020s, Mexico seeks to siphon high-value container traffic away from Panama, offering shippers a faster and more reliable east–west connection that bypasses both the canal's drought vulnerability and its draft limitations. The CIIT represents a deliberate attempt to engineer optionality through land-based infrastructure, illustrating how states respond to the fragility of maritime chokepoints by building parallel physical pathways.

This economic centrality is now entangled with great-power competition. The United States has historically exercised decisive influence over the canal, first through direct control and later through the 1977 treaties that transferred sovereignty to Panama while preserving certain security interests. In recent years, Chinese investment in regional port infrastructure—most notably the $1.3 billion Chancay deep-water megaport in Peru, inaugurated in late 2024—has introduced a new competitive dynamic. Chancay is designed to handle vessels too large for the Panama Canal's locks and aims to cut transit times between China and South America's Pacific coast by up to twenty days. The project illustrates how states are actively constructing alternatives to established chokepoints, seeking to reduce exposure to corridors whose governance or reliability they cannot fully control. Panama thus sits at the intersection of legacy strategic interests and emerging efforts to reconfigure trans-Pacific trade flows.

Governance, Sovereignty, and Parallel Human Pressures

The governance regime of the Panama Canal rests on the 1977 treaties and the subsequent establishment of the Panama Canal Authority as an autonomous agency of the Panamanian state. This framework has delivered relative stability and professional management, yet it also embeds the canal within the broader politics of Panamanian sovereignty and the enduring shadow of U.S. regional influence. The canal's operation requires continuous international cooperation—particularly in water management, security, and the facilitation of global trade—while remaining subject to domestic political pressures and the fiscal needs of the Panamanian state. The canal is therefore governed through a hybrid logic: formally autonomous, internationally embedded, and politically exposed.

Parallel to the canal's commercial function, Panama hosts one of the most intense human chokepoints on Earth: the Darién Gap. Hundreds of thousands of migrants traverse this dense jungle corridor each year, moving from South America toward the United States. This migration flow generates immense political friction, forcing the Panamanian government to divert security resources, manage humanitarian pressures, and navigate sustained diplomatic demands from Washington. The presence of this parallel human chokepoint demonstrates how demographic flows can destabilize the host state's governance capacity even as it seeks to maintain the commercial reliability of its primary maritime asset. Demographic pressures do not merely accompany chokepoint dynamics; they can actively complicate the state's ability to secure and operate the corridor itself.

The canal's hybrid exposure has become increasingly visible. Expanding digital management systems, traffic-control infrastructure, and port-adjacent logistics networks create new surfaces for cyber and hybrid threats. At the same time, the canal's freshwater dependency and the ecological pressures on the surrounding watershed link its operational resilience to environmental governance. These intersecting vulnerabilities—physical, digital, and ecological—illustrate the multi-domain character of contemporary chokepoint risk. They also highlight the particular challenges facing engineered corridors whose functionality depends on sustained investment, environmental stewardship, and institutional capacity.

The Panama Canal exemplifies a configuration of the Straits Systems Architecture defined by engineered compression, climate sensitivity, and strategic competition. Its significance derives from its role as the principal maritime link between the Atlantic and Pacific basins of the Americas. Its vulnerabilities are shaped by hydrological constraints that climate change is intensifying, by the concentration of global trade flows through a single managed corridor, and by the intersection of legacy geopolitical interests with emerging efforts to construct alternative pathways. By the mid-2020s, the canal's geophysical determinants were dominated by an artificial freshwater system whose reliability is increasingly tested by rainfall variability and domestic conflicts over watershed protection. Its strategic-military geometry reflects a legacy of U.S. influence now complicated by Chinese port investments on the Pacific coast of South America and Mexican efforts to build competing land bridges. Its economic and bio-industrial flows encompass critical container, grain, and energy movements whose disruption generates widespread delay and cost across the Western Hemisphere and beyond, while states actively construct rail alternatives to reduce exposure. Its governance regime rests on a stable but politically embedded treaty framework that must accommodate both international usage and domestic Panamanian priorities amid parallel migration pressures through the Darién Gap. Its demographic and civilizational interfaces include the canal's historical role in shaping Panamanian national identity, its ongoing function as a site of labor and transnational commerce, and the destabilizing effects of large-scale migration flows that divert state resources and generate external diplomatic friction. And its technological infrastructures integrate lock operations, traffic management, and expanding digital systems whose hybrid protection requirements are growing in parallel with the canal's commercial importance.

The canal therefore stands as both a testament to the transformative power of engineered infrastructure and a warning about the concentrated risks such infrastructure can create. Its future will be shaped by the interaction of climate trends, technological adaptation, domestic ecological governance, and the strategic choices of states seeking to secure or diversify their access to interoceanic movement. In this sense, Panama remains a defining node in the global chokepoint system—one whose engineered character makes its vulnerabilities both distinctive and instructive for understanding the broader dynamics of compression, dependency, and resilience in the twenty-first century.

This chapter completes the applied analysis of the world's principal maritime chokepoints. The analysis now turns to the synthetic and philosophical dimensions of the global system, where the accumulated insights of the preceding chapters are drawn together into a unified theoretical framework.

16

Chapter 16

The Global Chokepoint System

The world's straits do not operate in isolation. Each narrow corridor has its own geometry, governance, and strategic logic, but together they form a global chokepoint system—an interconnected architecture through which the world's economic metabolism, military mobility, and civilizational interactions flow. This system is not a collection of discrete passages; it is a structural network whose stability underpins the functioning of the international order. The vulnerabilities of one strait can propagate through others, and the leverage embedded in one corridor can reshape the strategic behavior of states across multiple regions. Understanding this system requires moving beyond regional analysis to a global perspective.

Interdependence, Asymmetry, and Non-Linearity as System Properties

The global chokepoint system is defined by three structural properties: interdependence, asymmetry, and non-linearity. These properties interact within the six subsystems of the Straits Systems Architecture, producing a dynamic network that is more than the sum of its parts.

Interdependence is the foundational property. Flows through one strait directly influence flows through others. The 2026 effective closure of the Strait of Hormuz, which triggered production shut-ins of 7.5–9.1 million barrels per day, forced Asian importers to seek alternative supplies, increasing pressure on the Strait of Malacca and accelerating rerouting through the Lombok and Sunda alternatives. Simultaneously, the ongoing Red Sea crisis (2023–2026) reduced Bab el-Mandeb and Suez transits by 57–62 percent, driving container traffic around the Cape of Good Hope and adding secondary pressure on the Danish Straits for LNG imports into Northern Europe. Even distant corridors are linked: heightened tensions in the Taiwan Strait in 2025–2026 disrupted semiconductor supply chains, which in turn affected manufacturing schedules in Europe and increased demand for just-in-time shipping through Gibraltar and the Danish Straits. In a Taiwan-contingency scenario, Malacca's exposure becomes both physical and financial: even absent closure, altered threat perceptions can reprice insurance, shift routing behavior, and propagate delay costs through global supply chains. These are not coincidental linkages; they are structural. The global economy is organized around a limited number of narrow corridors, so disruption in one forces reconfiguration across the network—often at significant cost in time, fuel, and insurance. **Market adaptation does not eliminate chokepoint risk; it redistributes cost into time, inflation, and ecological externalities.**

Asymmetry magnifies the strategic significance of these linkages. States do not depend on chokepoints equally. East Asian economies rely heavily on the Strait of Malacca and the Taiwan Strait for energy and semiconductors. European states depend on Suez, Gibraltar, and the Danish Straits for trade and the post-2022 energy transition. Middle Eastern producers rely on Hormuz, while Russia depends on the Bosporus and Dardanelles for grain and energy exports. The United States and its partners maintain global reach through multiple corridors, whereas China's dependency is more concentrated in the Indo-Pacific. The Bering Strait's emergence creates new asymmetries for Arctic actors. This uneven distribution creates structural leverage: the ability of one actor to influence others by controlling, threatening, or stabilizing a chokepoint. Asymmetry is therefore a source of both vulnerability and power.

Non-linearity is the property that makes the global chokepoint system inherently fragile. Small disruptions can produce large effects. The 2021 *Ever Given* grounding in Suez cost an estimated $9–10 billion per week; the 2023–2026 Red Sea crisis amplified this dynamic, with global ton-miles increasing by about 4 percent and freight rates on Asia–Europe routes surging dramatically. A single incident in the Taiwan Strait—whether a blockade or gray-zone escalation—could halt advanced semiconductor flows that underpin AI, defense, and automotive industries worldwide. Non-linearity transforms chokepoints into systemic risks precisely because compression concentrates impact.

The Six Subsystems at Global Scale

Africa as the Shock Absorber of the Chokepoint System (The Cape as a "Negative Chokepoint")

Africa's role in the global chokepoint system is often described only through exposure—migration, piracy, and environmental risk. Yet at system scale, Africa also functions as the primary shock absorber for chokepoint failure elsewhere. When Bab el-Mandeb and Suez become unsafe, the "alternative route" is not another narrow passage but the long oceanic circuit around the Cape of Good Hope. This produces what can be called a *negative chokepoint*: a place whose strategic significance does not arise from forced compression into a narrow channel, but from the way rerouting concentrates secondary burdens—ton-miles, emissions, crew risk, insurance premiums, and port congestion—onto a specific geographic arc. In a zero-slack fleet environment, the Cape is not merely longer; it is capacity-consuming, and therefore system-shaping.

The Cape rerouting logic converts disruption into time, and time into scarcity. Longer voyages tie up hulls and crews, reduce effective fleet capacity, and shift congestion into alternative ports and anchorages. It also transfers ecological and security risk toward African coastal waters: more vessel-days at sea increase emissions and fuel burn; more bunkering and maintenance demand concentrates in southern African ports; and a larger proportion of global shipping is pushed through waters where search-and-rescue capacity, spill response infrastructure, and maritime policing vary widely. In practical terms, the global system externalizes part of its resilience cost onto African littorals: the world reroutes, but Africa absorbs the ton-miles.

The Mozambique Channel as an emergent corridor. As Cape routing normalizes under stress, the Mozambique Channel becomes a critical leg in the reconfigured geometry of global shipping—especially for Asia–Europe and Gulf–Europe voyages diverted away from the Red Sea. In SSA terms, the channel's significance is produced by interaction effects: geophysical determinants (weather, currents, cyclone risk) combine with economic flows (container and energy rerouting), governance capacity (regional maritime security and port infrastructure), and strategic-military geometry (naval presence and piracy deterrence) to determine whether the "alternative" route is merely longer or systemically riskier. The key point is that optionality is not free: when the main corridor fails, the substitute corridor inherits a new density of exposure.

West African energy corridors and the Gulf of Guinea. Africa's chokepoint role is also visible in the Atlantic energy system. As European energy sourcing diversifies, West African oil and LNG routes—linking Nigeria, Angola, Equatorial Guinea, and emerging producers to European regasification and refinery networks—carry heightened strategic weight. Yet these flows move through a security environment where piracy, armed robbery, and offshore criminal networks have historically driven high insurance premiums and escort demand. Under SSA, this is a classic hybrid exposure: the economic subsystem (energy flows) interacts with governance and strategic-military geometry (coastal enforcement capacity and regional naval cooperation) to determine whether supply diversification produces resilience or simply shifts vulnerability from one corridor to another.

Treating Africa only as a sink for externalized risk misses a second reality: African coastal states can exercise agency in the chokepoint system. Port service capacity, bunkering, pilotage, and enforcement posture shape the practical viability of rerouting corridors. Port state control, environmental compliance checks, and sanctions enforcement can influence shadow shipping and substandard vessels seeking refuge in less regulated spaces. Regional security arrangements—whether ad hoc patrol coordination or broader maritime-domain-awareness initiatives—can shift the insurance and risk pricing that determines what routes remain commercially usable. In this sense, Africa is not only where the system's stress is absorbed; it is also where the system's adaptive capacity can be built or withheld.

Seeing Africa as a shock absorber therefore clarifies a central systems lesson: "alternatives" are themselves corridors with their own governance limits, ecological liabilities, and security geometries. Interdependence does not end when ships reroute; it relocates pressure into new places, often far from the initiating strait.

These three properties—interdependence, asymmetry, and non-linearity—operate across the six subsystems of the SSA model at the global scale. Geophysical changes (Arctic ice melt) open new corridors in the Bering Strait while altering optionality elsewhere. Military geometry in one theater (Taiwan) influences force posture in others (Bosporus). Economic flows cascade: a Hormuz shock ripples through Malacca and Suez. Governance regimes interact through UNCLOS and bilateral arrangements, yet remain fragmented. Civilizational interfaces shape narratives that cross regions, with identity-driven escalation risks in the Taiwan Strait influencing broader Indo-Pacific alignments. Technological infrastructures—undersea cables, satellite networks, and algorithmic logistics—create hybrid vulnerabilities that span multiple straits simultaneously.

CHAPTER 16

Transformation and the Reconfiguration of Global Order

In the modern global system, marine insurance syndicates function less as passive absorbers of geopolitical risk and more as de facto maritime regulators. During the 2024–2025 Red Sea and Black Sea crises, the global marine insurance market—nearly $40 billion in 2024—did not simply raise premiums; it determined which voyages were financeable at all. The rise of InsurTech has further strengthened this regulatory power. Underwriters now use artificial intelligence to identify high-risk ports, adjust rates dynamically, or withdraw coverage in near real time. When major insurers pull Protection and Indemnity or war-risk coverage from a corridor, that strait can become commercially non-viable even without formal closure. Market governance can therefore achieve blockade-like effects, enforcing or denying access through coverage decisions that operate outside traditional state authority. This capacity to shape global mobility through financial instruments is a powerful form of asymmetry embedded within the chokepoint system.

The global chokepoint system is now entering a period of profound transformation. Climate change is altering navigability in the Arctic, elevating the strategic importance of the Bering Strait. The energy transition is reshaping flows through Hormuz, Malacca, and the Danish Straits. Great-power competition is intensifying pressure on the Taiwan Strait and the Bosporus. Digital infrastructures—cables, satellites, autonomous systems—are creating new dependencies and new vulnerabilities across every corridor. These forces will not merely modify the system; they will reconfigure it. The architecture of global order will increasingly be defined by the stability, governance, and technological resilience of the world's narrow places.

This chapter synthesizes the applied analysis of individual straits into a unified model of the global chokepoint system. The book now turns to the next part, where the future of this system is examined: the emerging risks, the shifting patterns of dependency, and the strategic transformations that will define the century ahead.

17

Chapter 17

Future Shock: Climate, Conflict, and the New Chokepoint Era

The global chokepoint system is entering a period of accelerated transformation. Climate change, technological disruption, demographic pressure, and great-power competition are converging to reshape the strategic logic of the world's narrow places. These forces are not incremental; they are structural. They are altering the geophysical foundations of straits, the economic flows that pass through them, the military geometries that define their stability, and the governance regimes that regulate their use. The result is a new chokepoint era—one defined by volatility, asymmetry, and systemic exposure.

The 2026 Hormuz crisis offered a live demonstration. Production shut-ins of 7.5–9.1 million barrels per day cascaded through Malacca, Suez, and global energy markets within weeks, validating the nonlinear risks outlined in the preceding chapters. A single strait's disruption propagated across continents, revealing how tightly coupled the global system has become.

Climate Change: The Geophysical Reordering of Straits

Climate change is altering the physical environment of straits in ways that were once unthinkable. Rising sea levels threaten low-lying chokepoints such as the Malacca Strait and the Nile Delta approaches to Suez, where even modest increases in water depth and storm intensity complicate dredging and traffic management. Shifting monsoon patterns and more violent typhoons increase navigational risk in the Taiwan Strait, the Bab el-Mandeb, and the Gulf of Aden.

Most consequential is the transformation of the Arctic. The melting of sea ice is turning the Bering Strait from a marginal corridor into a strategic gateway. The Northern Sea Route carried **37.02 million metric tons** of cargo in 2025, while record-low winter sea-ice conditions into early 2026 extended the navigable season and reduced ice hazards. Climate change is not merely modifying the operational environment of straits; it is reconfiguring the global map of navigability, creating new corridors and exposing new vulnerabilities.

These geophysical shifts also concentrate ecological externalities. Longer navigable seasons and higher traffic volumes in previously ice-covered waters increase the risk of ship strikes on marine mammals, underwater noise pollution, and fuel spills in regions with limited response capacity. Environmental fragility becomes a structural feature of the new chokepoint era.

Climate policy itself now functions as a form of regulatory blockade. The European Union's extension of the Emissions

Trading System to maritime shipping from 2024—phased in through 2027—alongside the FuelEU Maritime regulation, has altered the mathematics of routing. Carbon pricing and well-to-wake fuel-intensity standards incentivize avoidance behaviors, including shifting transshipment away from EU ports to minimize emissions exposure. This "green protectionism" reshapes trade flows and chokepoint dependency through regulatory incentives as consequential as conflict or physical disruption.

Great-Power Competition: The Militarization of Narrow Places

Great-power rivalry is intensifying pressure on the world's most consequential chokepoints. The Taiwan Strait has become the central fault line of U.S.–China competition, with PLA activity reaching record levels in 2025. The Bosporus and Dardanelles sit at the intersection of NATO–Russia tensions, where Montreux enforcement continues to regulate naval access amid Black Sea incidents. The Strait of Hormuz remains a focal point of Middle Eastern competition, with the 2026 crisis demonstrating how regional actors can leverage asymmetric capabilities for strategic effect.

As military capabilities evolve—including long-range precision weapons, autonomous systems, and AI-driven surveillance—the geometry of deterrence within straits becomes more complex. Escalation can occur rapidly and unpredictably. Chokepoints are no longer passive transit corridors; they are active arenas of strategic signaling and coercion.

The Energy Transition: Redistribution, Not Relief

The energy transition is reshaping the flows that give chokepoints their significance. As economies shift from hydrocarbons to renewable power and diversified LNG and hydrogen networks, the dependency structure of straits changes. The 2026 Hormuz crisis accelerated diversification behavior, including greater reliance on Northern European LNG routing and the maritime logistics of offshore wind expansion.

The Strait of Malacca remains central to Asian energy flows, but the composition of those flows is diversifying. The Danish Straits have gained importance as conduits for offshore wind components and the logistics of Europe's green transition. The energy transition does not eliminate chokepoint dependency; it redistributes it.

This redistribution extends to the raw-material foundation of decarbonization. Electrification depends on lithium, cobalt, nickel, and rare earth elements, with processing and refining capacity heavily concentrated in Asia—predominantly China. Malacca and the Taiwan Strait thus become conduits not only for fuels and finished goods but for the building blocks of the green economy. Upstream policy can create downstream chokepoints: Indonesia's 2020 ban on unprocessed nickel exports illustrates how resource nationalism propagates bottlenecks through maritime corridors linking mines, smelters, battery manufacturing, and end markets.

CHAPTER 17

Technological Acceleration: The Fusion of Physical and Digital Risk

Technological acceleration is creating new forms of vulnerability and new mechanisms of control. Autonomous vessels, AI-driven logistics, and digital maritime infrastructure increase efficiency but also introduce cyber-nautical risks. The IMO's roadmap for a MASS Code—non-mandatory in 2026, mandatory by 2032—signals that autonomy is moving from experimentation to regulation.

Undersea cables, satellite networks, and data conduits converge in narrow corridors, making straits critical nodes in the global information system. As compute infrastructure moves into nearshore, floating, and subsea deployments for cooling and latency, straits become substrates for fixed digital assets—raising sovereignty, liability, and critical-infrastructure security stakes.

Repeated cable incidents in the Red Sea and Gulf approaches from 2024 to 2026 have demonstrated the fragility of these systems. Seabed militarization and port digitization compound exposure: UUVs create new tools for gray-zone monitoring and sabotage, while smart-port OT systems expand the attack surface for ransomware and insider disruption. To mitigate these vulnerabilities, states are accelerating the deployment of LEO satellite networks that provide resilient, low-latency communications and positioning redundancy when subsea or terrestrial infrastructure is compromised.

Demographic and Civilizational Pressure: The Human Chokepoint

Demographic and civilizational pressures are reshaping the human landscape of straits. Migration flows across the Bab el-Mandeb, Gibraltar, and the Aegean remain intense, driven by conflict, economic inequality, and climate stress. In 2025, the International Organization for Migration recorded **7,667 people dead or missing** on migration routes worldwide, underscoring how narrow maritime passages concentrate humanitarian risk alongside trade and security risk.

Urbanization around straits such as Singapore, Istanbul, and Copenhagen intensifies demand for infrastructure, governance, and security. Civilizational narratives—whether in the Taiwan Strait, the Bosporus, or Hormuz—shape political identities and strategic behavior. Demography does not merely influence the social environment of straits; it alters the political stakes embedded within them.

A particularly acute human vulnerability lies in the global maritime labor force. Ships do not move without crews, and the industry faces growing constraints in recruitment, retention, and specialized training for increasingly automated and alternative-fuel vessels. During the Red Sea crisis, labor frameworks expanded the practical right of seafarers to refuse transit through designated high-risk areas and to demand repatriation—forcing reroutes regardless of corporate preference. In this sense, the ultimate chokepoint is human: if crews will not sail, the physical openness of a strait becomes irrelevant.

Future Shock: A System Under Strain

These drivers interact to produce future shock—a condition in which the pace of systemic change exceeds the capacity of institutions, markets, and states to adapt. The global chokepoint system is becoming more interconnected, more technologically dependent, and more exposed to disruption. The vulnerabilities of one strait can propagate through others with increasing speed. The leverage embedded in chokepoints is becoming more consequential, and the risks associated with their disruption more severe.

Emerging Dynamics of the New Chokepoint Era

Three dynamics will define the coming era.

First, new corridors will rise. Arctic routes, engineered passages, and digitally mediated routing systems will create new chokepoint dependencies. The Bering Strait's transformation is the clearest example, but similar dynamics will emerge in Malacca bypass routes and in hydrogen and ammonia shipping lanes.

Second, supply chains will become politicized. States will increasingly treat chokepoints as instruments of national strategy rather than neutral corridors. Insurance, environmental compliance, and sanctions enforcement will function as gatekeeping tools capable of restricting access without kinetic action.

Third, physical and digital vulnerabilities will fuse. Cyber-nautical disruption, GNSS interference, cable cuts, and AI-enabled electronic warfare will create new escalation channels. Because global fleet capacity is finite, simultaneous disruptions generate zero-slack conditions: ships absorbed on longer reroutes are unavailable elsewhere, amplifying shortages and price spikes.

The Centrality of Chokepoints in Global Order

These dynamics will reshape the global order. The stability of the international system will depend not only on the governance of states but on the resilience of the world's narrow places. The chokepoint is becoming the central unit of strategic analysis—the point where climate, conflict, technology, and civilizational identity converge.

This chapter introduces the forces that will define the future of the global chokepoint system. The chapters that follow examine these forces in greater detail, mapping the trajectories of transformation and the strategic implications for states, markets, and global governance.

18

Chapter 18

The New Geography of Risk

The Five Transformations Reshaping Risk

The geography of global risk is shifting. For most of the modern era, strategic vulnerability was concentrated in a small number of predictable corridors—Hormuz, Malacca, Suez, Gibraltar, the Bosporus. These straits formed the backbone of the global system, and their stability or disruption shaped the behavior of states and markets. That geography is now being reconfigured. Climate change, technological acceleration, geopolitical realignment, and the restructuring of global supply chains are redistributing risk across regions, sectors, and systems. The result is a new geography of risk—one that is more diffuse, more dynamic, and more difficult to govern. The events of 2025–2026 have made this shift unmistakable: the Hormuz crisis produced immediate supply shocks felt in Malacca and the Danish Straits;

the ongoing Red Sea rerouting permanently altered ton-mile patterns; and record PLA activity in the Taiwan Strait elevated semiconductor risk to a global concern.

The first transformation in this geography is the northward shift of strategic relevance. As Arctic ice recedes, the Bering Strait and the Northern Sea Route are emerging as new corridors of global movement. The NSR carried 37.02 million metric tons of cargo in 2025, and vessel transits through the Bering Strait have risen sharply over the past decade as the navigable season lengthens. Marine Exchange of Alaska data shows transits increasing from 242 in 2010 to 665 in 2024, reflecting how Arctic access is turning environmental change into traffic growth. These routes are not yet reliable year-round, but their potential is altering strategic calculations. The Arctic is becoming a zone where environmental change intersects with great-power competition, resource extraction, and digital infrastructure. The geography of risk is expanding into regions that were once peripheral, creating new chokepoints and exposing new vulnerabilities. The Arctic is no longer a frontier; it is a strategic domain. These geophysical shifts also concentrate ecological externalities: longer navigable seasons and higher traffic volumes in previously ice-covered waters increase the risk of ship strikes on marine mammals, underwater noise pollution that disrupts indigenous subsistence practices, and potential fuel spills in regions with limited response infrastructure.

The second transformation is the eastward intensification of systemic exposure. The Indo-Pacific—anchored by the Taiwan Strait, the South China Sea, and the Strait of Malacca—has become the center of global economic activity and geopolitical tension. The region's chokepoints carry the world's most critical flows: semiconductors (Taiwan controls ~72 percent of pure-play foundry capacity), energy, manufactured goods, and digital infrastructure. The concentration of strategic value in these corridors creates a geography of risk that is dense and asymmetric. A disruption in the Taiwan Strait would reverberate through Malacca, Suez, and the Danish Straits. The Indo-Pacific is not merely a regional system; it is the core of the global chokepoint architecture.

The third transformation is the reemergence of Europe's maritime vulnerabilities. The Danish Straits, Gibraltar, and the Bosporus are becoming more consequential as Europe undergoes energy transition, military realignment, and geopolitical fragmentation. The war in Ukraine and the 2023–2026 Red Sea crisis have reconfigured flows of energy, grain, and industrial inputs, increasing pressure on the Bosporus and the Danish Straits for LNG and offshore wind logistics. The Mediterranean is once again a zone of migration pressure, civilizational tension, and strategic competition. Europe's geography of risk is no longer defined solely by land borders; it is increasingly maritime.

The fourth transformation is the expansion of digital chokepoints, which overlay and intersect with physical ones. Undersea cables, satellite networks, and data conduits converge in narrow maritime corridors. More than 1.4 million kilometers of submarine fiber-optic cables carry over 99 percent of global internet and financial data traffic, with major clusters in the Bab el-Mandeb/Red Sea, Hormuz, Malacca, and Gibraltar regions. Repeated cable cuts in 2024–2026 (including incidents in the Red Sea and Hormuz approaches) caused regional internet disruptions and latency spikes. The geography of risk is therefore no longer limited to physical space; it includes the invisible architecture of the digital world.

The fifth transformation is the fragmentation of global supply chains, which is redistributing risk across multiple corridors. States are diversifying trade routes, reshoring manufacturing, and building redundancy into critical industries. These shifts reduce dependency on some chokepoints while increasing pressure on others. The rise of India as a manufacturing hub increases the importance of the Indian Ocean and the Strait of Malacca. The diversification of energy flows increases the relevance of the Danish Straits and the Cape of Good Hope. Supply-chain fragmentation does not eliminate chokepoint risk; it redistributes it.

> Fragmentation does not dissolve concentration; it redistributes it into new corridors, new materials, and new domains.

Structural Features of the New Geography

These transformations interact to produce a new geography of risk characterized by three structural features:

- **Multipolar vulnerability.** Risk is no longer concentrated in a few predictable corridors; it is distributed across multiple regions and systems. The Bering Strait's emergence, the Indo-Pacific core, and Europe's reexposed flanks now operate in parallel.
- **Layered exposure.** Physical, digital, economic, and civilizational vulnerabilities overlap, creating hybrid chokepoints where disruption can propagate across domains. A cable cut

in the Red Sea or Hormuz can simultaneously affect data flows, energy markets, and migration dynamics.

- **Dynamic instability.** The geography of risk is fluid, shaped by climate patterns, technological shifts, and geopolitical realignments that evolve faster than governance structures can adapt. The SSA model reveals how changes in one subsystem (e.g., geophysical ice melt) cascade through military geometry, economic flows, and technological infrastructures.

The five transformations reshaping the global chokepoint system can be understood as a set of directional shifts—geographical, economic, technological, and strategic—that are redistributing risk and redefining the architecture of global mobility.

The first is a **northward shift**, driven by Arctic emergence. The Bering Strait and the Northern Sea Route are becoming seasonally reliable corridors as geophysical determinants change, strategic-military geometry intensifies, and technological infrastructures expand. A passage that once functioned as a marginal frontier is becoming strategically salient, yet its ecological fragility and limited response capacity mean that accidents or disruptions carry disproportionately severe consequences.

The second is an **eastward intensification** centered on the Indo-Pacific. The Taiwan Strait, the Strait of Malacca, and the South China Sea approaches now anchor the world's manufacturing and semiconductor dependencies. Here, strategic-military geometry, economic and bio-industrial flows, technological infrastructures, and civilizational interfaces converge. Regional coercion in this zone can rapidly escalate into global industrial disruption, making it the most consequential concentration of chokepoint risk on Earth.

The third transformation is a **Western Hemisphere rewiring**, as states construct new "immune infrastructure" to reduce exposure to legacy chokepoints. Trans-Pacific routes and South American hubs are being redesigned through megaproject ports, rail corridors, and alternative logistics chains. These shifts alter economic flows, governance regimes, and technological infrastructures, bypassing traditional leverage points while creating new hubs—and new dependencies—elsewhere.

The fourth is **Europe's maritime reexposure**. The Danish Straits, Gibraltar, the Bosporus, and the Suez approaches are being reshaped by the energy transition, hybrid threats, demographic pressures, and dense technological infrastructures. Maritime access is once again a central European security variable, as the continent's energy systems, industrial base, and political stability depend on the reliability of these corridors.

The fifth transformation is the rise of **digital and orbital chokepoints**. Cable clusters in the Bab el-Mandeb, the Red Sea, Hormuz, Malacca, and Gibraltar have become critical nodes in the global information system, while Low Earth Orbit constellations provide redundancy but introduce congestion and new governance challenges. Technological infrastructures and regulatory regimes now determine whether cyber-nautical disruption or cable cuts can impose closure-like effects. Risk is migrating into the vertical domain, where digital and orbital dependencies shape the stability of physical corridors.

Together, these shifts demonstrate that chokepoint risk is no longer confined to a small set of inherited passages. It is migrating into new geographies, stacking across domains, and becoming increasingly difficult to stabilize with legacy governance tools.

Strategic Implications and the Centrality of Chokepoints

This new geography challenges traditional strategic assumptions. It requires states to think in terms of systems rather than regions, interdependencies rather than isolated threats, and resilience rather than control. The chokepoint is no longer a static feature of geography; it is a dynamic node in a global architecture of risk.

This chapter maps the emerging contours of that architecture. The next chapter examines how states are responding—the strategies, doctrines, and institutional innovations that are shaping the future governance of the world's narrow places.

19

Chapter 19

Statecraft in the Chokepoint Century

Littoral Agency and the Reconfiguration of Power

A recurring blind spot in chokepoint analysis is the tendency to treat littoral states as passive geography—territories where risk accumulates rather than political actors who actively shape it. In reality, host states convert exposure into leverage through pricing, law, infrastructure, enforcement, and diplomatic brokerage. Their agency is not incidental; it is constitutive of how chokepoints function. Djibouti demonstrates brokerage under fragility: its influence derives not from unilateral control but from its ability to host a dense cluster of foreign bases and cable-landing infrastructure, transforming proximity to Bab el-Mandeb into revenue, security guarantees, and diplomatic centrality. Egypt represents the sovereign operator with fiscal exposure, managing Suez through a professionalized authority while remaining vulnerable to revenue shocks that threaten macroeconomic stability. Indonesia functions as an archipelagic gatekeeper whose control of Sunda and Lombok shapes the credibility of Malacca alternatives, influencing global risk pricing without ever closing the primary corridor. Morocco exercises southern-shore leverage at Gibraltar through port competition, migration enforcement, and maritime security cooperation, shaping the corridor's economic and political environment through logistical and demographic influence rather than overt military power. Across these cases, littoral agency emerges from the interaction of SSA subsystems: governance regimes provide the legal authority to condition access; economic flows supply rents and bargaining chips; strategic-military geometry determines whether leverage is exercised through enforcement or brokerage; demographic interfaces shape legitimacy and migration bargaining; and technological infrastructures define how much control is possible in practice. Treating host states as strategic actors rather than passive geography aligns with the book's core claim: chokepoint power is an interaction between structure and choice, and the chokepoint is becoming the central unit of strategic practice in an era defined by climate volatility, technological acceleration, and geopolitical fragmentation. As global flows become more brittle and more contested, the political value of littoral agency rises: the ability to stabilize, monetize, or weaponize a corridor becomes a form of statecraft that can elevate small states, constrain great powers, and reorder regional hierarchies.

The Six Dimensions of Chokepoint Statecraft

This emerging architecture of statecraft can be understood through six interlocking dimensions that together define how power is exercised in the chokepoint century. Positional strategy involves the deliberate cultivation of geographic leverage, increasingly expressed through "white hull diplomacy," where coast guards—not navies—become the primary instruments of practical control. China's jurisdictional normalization tactics in the Taiwan Strait and South China Sea exemplify this shift, forcing rivals into dilemmas of escalation and legitimacy. Infrastructural strategy uses ports, pipelines, cables, and digital systems to shape chokepoint dynamics, embedding structural power in the physical and informational architecture of global movement. Legal-institutional strategy deploys treaties, governance regimes, and lawfare to constrain adversaries, with Montreux, Copenhagen, UNCLOS interpretations, and emerging environmental rulings reshaping access and cost structures. Technological strategy integrates digital systems, autonomous platforms, and AI-driven surveillance into chokepoint governance, while simultaneously expanding hybrid vulnerabilities through UUVs, smart-port OT systems, and seabed militarization. Economic-industrial strategy restructures supply chains to manage chokepoint exposure, from semiconductor friend-shoring to Europe's post-2022 energy diversification and China's upstream control of critical minerals. A particularly acute human dimension lies in the maritime labor force, whose refusal to transit high-risk zones during the Red Sea crisis demonstrated that the ultimate chokepoint is human: if crews will not sail, physical openness becomes irrelevant. Finally, civilizational strategy uses identity, narrative, and historical memory to shape behavior around straits, influencing escalation thresholds in places such as the Taiwan Strait, the Bosporus, and Gibraltar. Together, these dimensions reflect a shift from territorial control to flow control, from dominance over land to influence over movement, and from static borders to dynamic corridors. States that master this grammar will shape the future of global order; those that fail will find themselves constrained by vulnerabilities they cannot manage. The chokepoint century demands a strategic imagination attuned to systems rather than regions, interdependencies rather than isolated threats, and resilience rather than dominance. It requires states to understand that power now resides not only in the ability to project force, but in the ability to shape the conditions under which movement occurs—who moves, what moves, how it moves, and under whose rules.

20

Chapter 20

The Architecture of Resilience

The chokepoint century demands a new strategic architecture—one that does not merely manage risk but absorbs, adapts to, and transforms it. Resilience is no longer a peripheral concern; it is the central organizing principle of global stability. Yet resilience is often misunderstood. It is not redundancy, not robustness, not the simple capacity to withstand disruption. Resilience is a systemic property: the ability of a complex architecture to maintain function under stress, reconfigure itself when necessary, and evolve in response to structural change. In the global chokepoint system, resilience must be designed, not assumed. It does not diminish chokepoint centrality; it increases the system's tolerance for chokepoint stress without eliminating dependency.

Resilience rests on six pillars. The first is **geophysical adaptation**—recognizing that climate change is altering the physical environment of straits in ways that require anticipatory design. Rising sea levels threaten low-lying corridors such as the Malacca Strait and the Nile Delta approaches to Suez, where even modest increases in water depth or storm intensity complicate dredging and traffic management. The Arctic is becoming navigable, creating new corridors and new vulnerabilities in the Bering Strait. Resilience therefore requires adaptive infrastructure: enhanced dredging, coastal fortification, real-time environmental monitoring, dynamic charting, and nature-based solutions. Wetlands, mangroves, and reef systems can attenuate wave energy, stabilize sediments, and reduce long-run maintenance burdens. Geophysical adaptation is not only about resisting nature but working with ecology to protect infrastructure.

The second pillar is **military stabilization**—the creation of deterrence architectures that reduce the likelihood of escalation in contested straits. Stabilization does not mean militarization; it means predictability. Confidence-building measures, deconfliction protocols, surveillance transparency, and crisis-management mechanisms reduce the risk of accidental escalation in the Taiwan Strait, the Bosporus, and the Strait of Hormuz. NATO consolidation in the Baltic after Finland and Sweden's accession, and the Danish Straits' role as a "NATO lake," demonstrate how alliance geometry can stabilize access. Resilience requires military geometry that remains stable under stress rather than brittle under pressure.

The third pillar is **economic diversification**—the restructuring of global supply chains to reduce chokepoint dependency. Diversification does not eliminate the importance of straits; it distributes risk across multiple corridors. Expanded LNG routing reduced pressure on Hormuz during the 2026 crisis. Distributed semiconductor manufacturing—accelerated by CHIPS Act investments in the United States, Europe, Japan, and India—lowers exposure to the Taiwan Strait. Expanded port capacity in the Mediterranean and North Sea mitigates reliance on Suez. This diversification extends to the raw-material foundation of the energy transition: lithium, cobalt, nickel, and rare earth elements, whose refining is heavily concentrated in Asia. Resource nationalism, such as Indonesia's ban on unprocessed nickel exports, creates upstream bottlenecks that propagate through the same corridors that carry fuels and manufactured goods. The ultimate defense against material chokepoints is the circular economy: scaling urban mining, e-waste recycling, and materials recovery to create secondary supply with dramatically lower emissions. Economic diversification is not a retreat from globalization; it is a redesign of its architecture.

A particularly acute human dimension of resilience lies in the **global maritime labor force**. Ships cannot function without crews, and the industry faces structural deficits—especially in officers trained for highly automated and alternative-fuel vessels. During the 2023–2026 Red Sea crisis, labor agreements affirmed seafarers' right to refuse transit into high-risk areas, forcing reroutes regardless of commercial preference. The ultimate chokepoint is human: if crews will not sail, the physical openness of a strait becomes irrelevant.

The fourth pillar is **governance modernization**—the creation of institutional frameworks capable of managing hybrid chokepoint environments. Traditional maritime law is insufficient for a world in which cyber-nautical disruption, autonomous vessels, and digital infrastructure shape chokepoint dynamics. Governance must evolve to incorporate data integrity, cyber defense, environmental protection, and shared surveillance.

The Danish Straits demonstrate the stabilizing power of institutionalized cooperation; the Bab el-Mandeb demonstrates the risks of its absence. States are also expanding environmental and insurance-based regulatory tools—such as P&I coverage verification and emissions compliance—to manage liability and restrict access without suspending passage rights. The 2024 ITLOS advisory opinion treating greenhouse gas emissions as marine pollution under UNCLOS reinforces the use of environmental standards as governance instruments.

The fifth pillar is **civilizational accommodation**—recognizing that identity, narrative, and historical memory shape the stability of straits as much as military or economic factors. The Taiwan Strait is a civilizational boundary. The Bosporus is embedded in the legacy of empire. Gibraltar symbolizes sovereignty and continuity. Resilience requires political architectures that acknowledge these narratives rather than suppress them. Civilizational accommodation is not concession; it is strategic realism that reduces the risk of identity-driven escalation.

The sixth pillar is **technological redundancy and integrity**—the protection of digital infrastructure that underpins maritime movement. Undersea cables, satellite networks, port automation systems, and AI-driven logistics are now integral to chokepoint stability. Their failure can halt global flows as effectively as a physical blockade. Resilience requires encryption, physical redundancy, cyber defense, and diversified pathways across subsea, terrestrial, and space layers. Repeated cable incidents in the Red Sea and Hormuz approaches from 2024 to 2026 underscore the urgency of hybrid protection. UUVs create new sabotage possibilities, while smart-port operational technology expands the attack surface. LEO satellite constellations provide redundancy but share vulnerabilities with GNSS under jamming or spoofing. For this reason, resilience planning increasingly emphasizes alternative positioning, navigation, and timing systems—such as terrestrial eLoran—so that a space-layer failure does not become a single point of systemic collapse.

Together, these pillars form the **architecture of resilience**—a multi-layered system integrating physical, military, economic, institutional, civilizational, and technological dimensions. This architecture is not defensive; it is strategic design. It transforms chokepoints from sources of systemic fragility into components of systemic stability. Distributed dependency ensures that no single corridor can accumulate disproportionate leverage. Adaptive governance evolves in step with environmental and technological change, adjusting institutions, rules, and enforcement as the operating environment shifts. Hybrid security integrates physical and digital protection into a single architecture, recognizing that ports, cables, satellites, and ships now form one interdependent system. These principles do not eliminate risk; they shape it. They do not prevent disruption; they prevent disruption from cascading into systemic collapse. They do not guarantee stability; they create the conditions under which stability can be sustained even as the world's narrow places become more contested, more complex, and more essential to global order.

21

Chapter 21

Operationalizing Resilience

Resilience becomes meaningful only when it is operationalized. Conceptual frameworks provide structure, but stability in the chokepoint century depends on the creation of mechanisms—technical, institutional, economic, and strategic—that translate design into practice. The architecture of resilience outlined in the previous chapter must therefore be rendered into operational systems capable of absorbing disruption, redistributing risk, and sustaining global flows under conditions of stress. Operationalizing resilience is not a matter of policy preference; it is a structural requirement of the emerging global order.

The Seven Mechanisms of Operational Resilience

The first operational mechanism is **multi-corridor redundancy**—the deliberate creation of alternative pathways for critical flows. This includes expanding port capacity, diversifying shipping routes, and investing in overland and digital corridors that reduce dependency on single straits. The India–Middle East–Europe Economic Corridor and the accelerated development of Arctic routes through the Bering Strait exemplify this approach, while secondary bypasses around Malacca—Lombok and Sunda—saw increased utilization during the 2023–2026 Red Sea crisis. Multi-corridor redundancy does not eliminate chokepoints; it reduces the systemic consequences of their disruption. It transforms vulnerability into optionality.

The second mechanism is **real-time situational awareness**, enabled by integrated surveillance, data fusion, and predictive analytics. Maritime domain awareness systems, satellite constellations, autonomous sensors, and AI-driven risk models allow states and institutions to detect anomalies, anticipate disruptions, and coordinate responses. This capability proved decisive during the 2026 Hormuz crisis and ongoing Red Sea operations, where commercial platforms such as Windward and Planet enabled rapid rerouting decisions. Real-time awareness is not merely informational; it is strategic. It provides the temporal advantage necessary to prevent local incidents from becoming systemic shocks.

The third mechanism is **distributed logistical capacity**—the decentralization of storage, processing, and manufacturing nodes across regions. Distributed energy storage and LNG terminals have reduced pressure on Hormuz; distributed semiconductor fabrication, accelerated by CHIPS Act investments in the United States, Europe, Japan, and India, is lowering exposure to the Taiwan Strait; distributed grain reserves have mitigated risks associated with Bosporus disruptions. This mechanism transforms supply chains from linear systems into networked architectures capable of absorbing localized failures. Distributed capacity is not redundancy; it is structural resilience.

The fourth mechanism is **institutional interoperability**, which enables states and organizations to coordinate across jurisdictions, sectors, and domains. This includes harmonized maritime regulations, shared crisis-response protocols, joint surveillance platforms, and integrated cyber-nautical defense systems. The Danish Straits demonstrate the stabilizing power of interoperable governance through the Copenhagen Treaty, EU law, and NATO frameworks; the Bab el-Mandeb demonstrates the risks of its absence. Institutional interoperability transforms fragmented governance into a cohesive system capable of managing hybrid threats.

The fifth mechanism is **strategic buffering**, the deliberate creation of temporal and material margins that absorb shocks. Strategic petroleum reserves, semiconductor stockpiles, diversified food storage, and flexible shipping schedules provide buffers that prevent immediate systemic collapse during disruptions. Buffering is not inefficiency; it is insurance. It provides the temporal space necessary for adaptive response.

The sixth mechanism is **hybrid infrastructure protection**, which integrates physical security, cyber defense, and digital integrity into a unified operational framework. Undersea cables, port automation systems, satellite networks, and maritime communication platforms must be protected as a single system. Hybrid protection requires encryption, physical redundancy, intrusion detection, and coordinated cyber-physical response capabilities. Repeated cable incidents in the Red Sea and Hormuz approaches in 2024–2026 have underscored the urgency of this approach. In the chokepoint century, the failure of digital infrastructure can halt global flows as effectively as a physical blockade.

The seventh mechanism is **adaptive governance**, which enables institutions to evolve in response to environmental, technological, and geopolitical change. Adaptive governance incorporates scenario planning, stress testing, regulatory flexibility, and iterative policy design. It recognizes that chokepoint dynamics are fluid and that governance must evolve at the pace of systemic transformation. Adaptive governance is not reactive; it is anticipatory.

Together, these mechanisms operationalize the architecture of resilience. They translate conceptual pillars into actionable systems. They create a global environment in which chokepoints remain critical but no longer function as single points of systemic failure. Operational resilience is therefore not a defensive posture; it is a strategic capability. It enables states and institutions to navigate a world defined by volatility without succumbing to it.

Strategic Outcomes of the Resilience Architecture

The operationalization of resilience produces three strategic outcomes. **Reduced systemic fragility** ensures that disruptions in one corridor no longer propagate uncontrollably across the global system; redundancy and buffering reduce the probability of cascade failure under polycrisis conditions. **Enhanced strategic autonomy** gives states the capacity to manage chokepoint exposure without relying on unilateral dominance; diversification, closed-loop industrial design, and sovereign backstops reduce coercive leverage. **Increased global stability** emerges when adaptive governance, hybrid protection, interoperability, and enforcement tools—including port-state control—create conditions in which volatility is absorbed rather than amplified.

From Conceptual Design to Systemic Capability

This chapter establishes the operational foundation for resilience in the chokepoint century. The next chapter turns to the final layer of the book's architecture: the philosophical and civilizational implications of a world organized around narrow places—and what it means for the future of global order.

22

Chapter 22

The Philosophy of Narrow Places

The chokepoint is not only a strategic object; it is a philosophical form. It reveals how systems behave under constraint, how power accumulates in spaces of compression, and how civilizations understand the relationship between openness and limitation. Narrow places are not anomalies in the global order; they are expressions of a deeper structural truth: all complex systems depend on points of concentration where flows converge, decisions crystallize, and vulnerabilities become visible. The philosophy of narrow places is therefore a philosophy of structure—an inquiry into the conditions under which global order becomes possible, fragile, and transformable. The events of 2025–2026, from the Hormuz crisis to record PLA activity in the Taiwan Strait and accelerating Arctic transits through the Bering Strait, have rendered this philosophy empirically urgent rather than abstract.

CHAPTER 22

The Six Insights of Narrow Places

The first philosophical insight is the **primacy of constraint**. Modern political thought often imagines freedom as the natural state of systems and constraint as an imposition. Chokepoints invert this assumption. They demonstrate that constraint is foundational: it shapes flows, organizes behavior, and creates the conditions under which order emerges. Malacca channels the economic metabolism of Asia; the Bosporus regulates the naval geometry of the Black Sea; the Taiwan Strait structures the strategic imagination of the Indo-Pacific. Constraint is not the opposite of freedom; it is the architecture within which freedom becomes meaningful. This insight dismantles the early-2000s belief that globalization had created a "flat," borderless world. Network topology reveals a different reality: global trade behaves like a scale-free network, where preferential attachment concentrates flows into ultra-dense hubs. Globalization did not flatten the world—it hyper-centralized it. Chokepoints are not exceptions; they are structural outcomes of optimization, where efficiency produces concentration and concentration produces fragility.

The second insight is the **asymmetry of influence**. Narrow places reveal that power is not evenly distributed across space. It accumulates where flows compress, where access is limited, and where the cost of disruption is high. This asymmetry is not merely geopolitical; it is ontological. A single vessel in Suez can immobilize global trade. A single miscalculation in the Taiwan Strait can destabilize the international order. A governance failure in the Bab el-Mandeb can reshape maritime insurance markets. Systems are defined not by their averages but by their critical points. This asymmetry also produces an ethical inequity: straits operate as global commons for the benefit of international trade, yet the risks are localized. When an underinsured shadow tanker spills oil in the Danish Straits, or when a ship sinks in the Bab el-Mandeb, the global system reroutes and reprices—but the host state inherits poisoned fisheries, devastated ecology, and massive cleanup liabilities. This is the tragedy of the chokepoint commons: the world demands frictionless passage while privatizing catastrophic risk onto littoral communities.

The third insight is the **interdependence of vulnerability and leverage**. Narrow places are simultaneously sources of risk and instruments of power. They expose states to disruption while enabling them to influence others. This duality is structural, not contradictory. The same corridor that sustains a state's economy can be used to coerce it. The same strait that enables global trade can become a site of blockade. The same digital infrastructure that accelerates maritime logistics can be exploited for cyber-nautical disruption. Vulnerability and leverage are two expressions of the same property: dependency. The SSA model makes this interdependence visible across all six subsystems. In the contemporary era, this duality is amplified by the bifurcation of global shipping into parallel "white" and "dark" logistical systems, where shadow fleets operate under incompatible legal and financial realities within the same physical straits.

The fourth insight is the **temporality of systems**. Chokepoints reveal that global order is not static; it is temporal. Flows move, crises escalate, governance evolves, and technologies transform. The significance of a strait is not fixed by geography alone; it is shaped by the rhythms of climate, conflict, and economic change. The Bering Strait was once marginal; it is now emerging. Hormuz was once peripheral; it became central in 2026. The Danish Straits were once contested; they are now institutionalized. The philosophy of narrow places is therefore a philosophy of time—an inquiry into how systems evolve under pressure. Today's landscape constitutes a polycrisis in which risks from distinct domains unfold simultaneously and compound one another. Because the global merchant fleet's capacity is finite, simultaneous disruptions generate zero-slack conditions: ships absorbed on extended reroutes are unavailable elsewhere, amplifying shortages and price spikes across the system.

The fifth insight is the **civilizational dimension of space**. Narrow places are not merely physical corridors; they are symbolic thresholds. They mark boundaries between cultural spheres, historical narratives, and political identities. The Bosporus separates continents and civilizations. Gibraltar stands at the hinge of the Mediterranean world. The Taiwan Strait embodies competing visions of political order. These meanings shape strategic behavior as much as military capability or economic dependency. A particularly acute human dimension lies in the maritime labor force: if crews refuse to sail into high-risk zones, or if shortages of qualified officers deepen, the physical openness of a strait becomes irrelevant. Migration flows across the Bab el-Mandeb, Gibraltar, and the Aegean further reshape the social environment of straits, transmitting labor, humanitarian pressure, and security concerns into littoral politics.

The sixth insight is the **hybridity of modern vulnerability**. Narrow places are no longer purely physical. They are hybrid spaces where digital, economic, ecological, and civilizational systems intersect. Undersea cables, satellite networks, and AI-driven logistics overlay the physical geometry of straits.

Climate change alters navigability. Migration reshapes the social environment. Cyber-nautical disruption introduces new forms of risk. The chokepoint becomes a microcosm of the global system: a space where multiple layers of reality converge into a single point of exposure. This hybridity exposes a deep ontological clash. Capital, data, and algorithmic routing models desire to move at the speed of light—they seek zero friction. Yet they are forced to manifest in physical reality: 400-meter-long steel ships navigating 700-meter-wide channels of saltwater against currents and weather. Clausewitz wrote about "friction" in war; straits are generators of friction in globalization. The clash between digital instantaneity and geophysical drag cannot be resolved—only managed. Port authorities respond with digital twins for predictive stress-testing, while true redundancy increasingly requires backing up space-based systems with hardened terrestrial alternatives.

The Structural Logic of Compression

Together, these insights form the philosophy of narrow places—a conceptual framework that reveals the deeper logic of the global chokepoint system. This philosophy is not abstract; it is structural. It explains why chokepoints matter, why they endure, and why they shape the evolution of global order. It shows that narrow places are not exceptions to the system; they are expressions of its fundamental architecture. The shift toward circular economies represents civilization's attempt to engineer its way out of geographic determinism. Port State Control represents the reassertion of sovereign boundaries over the illusion of the free seas. Sovereign insurance backstops represent the state's assumption of ultimate financial responsibility when private markets abandon the commons. These are not merely operational tools; they are philosophical statements about the relationship between sovereignty, risk, and global order.

The philosophy of narrow places leads to a final insight: global order is always built on compression. It is the narrow places—not the open oceans—that reveal the structure of the world. They show where power accumulates, where vulnerability concentrates, and where the future of the international system will be decided. The chokepoint is the point at which the abstract architecture of globalization becomes concrete—where the mathematical inevitability of concentration meets the lived reality of constraint, friction, and consequence. In this sense, the philosophy of narrow places is not a marginal inquiry; it is a reflection on the conditions of order itself.

This chapter completes the conceptual arc of the book. The final chapter turns to synthesis—the articulation of a unified theory of chokepoints and the implications for the future of global governance.

23

Chapter 23

Toward a Unified Theory of Chokepoints

A unified theory of chokepoints must account for the full complexity of the global system: its physical constraints, its strategic behaviors, its economic dependencies, its civilizational narratives, and its technological infrastructures. It must explain why narrow places matter, how they shape global order, and why their significance persists across eras of transformation. Such a theory cannot be merely descriptive; it must be structural. It must reveal the deep logic that binds the world's straits into a coherent architecture of power, vulnerability, and systemic evolution. The 2025–2026 crises—from Hormuz's effective closure and production shut-ins of 7.5–9.1 million barrels per day to record PLA activity in the Taiwan Strait and accelerating Arctic transits through the Bering Strait—have provided empirical validation for this unified view.

The Six Principles of the Unified Theory

1. Structural Compression

Chokepoints exist because global flows—of energy, goods, data, people, and military power—must pass through narrow corridors where geography, infrastructure, or governance compress movement. Compression creates predictability, but it also creates exposure. It concentrates risk, amplifies disruption, and generates leverage. The 2026 Hormuz crisis demonstrated this with brutal clarity: a navigable channel only a few kilometers wide disrupted nearly one-quarter of global seaborne oil. Global trade behaves as a scale-free network driven by preferential attachment, meaning many links depend on a small number of ultra-concentrated hubs. Chokepoints are not anomalies; they are mathematical inevitabilities in any optimized system. Globalization did not flatten the world—it hyper-centralized it. Structural compression is therefore the foundational condition of chokepoint significance.

2. Dependency Asymmetry

Not all actors rely on chokepoints equally. Some depend on them for survival; others for advantage. This asymmetry creates a distribution of vulnerability and leverage that shapes strategic behavior. East Asia's concentrated reliance on Malacca and the Taiwan Strait, contrasted with more diversified European and American options, illustrates this principle. The ethical inequity is profound: straits operate as global commons for the benefit of international trade, yet the risks are aggressively localized. When a shadow-fleet tanker spills oil in the Danish

Straits or a ship sinks in the Bab el-Mandeb, the global market reroutes and reprices—but the host state inherits devastated ecology and massive cleanup costs. This is the tragedy of the chokepoint commons. Dependency asymmetry also maps onto a geopolitical divide: the world's most consequential physical chokepoints lie in the Global South, while the financial architecture that governs movement—insurance clubs, currencies, banking—remains concentrated in the Global North. Systems that distribute benefit globally and harm locally are analytically unstable; they invite backlash, non-compliance, and legitimacy crisis.

3. Systemic Interdependence

Chokepoints do not operate in isolation; they form a network. Disruption in one corridor alters flows through others. A crisis in the Taiwan Strait affects Malacca, Suez, and the Danish Straits. A blockage in Suez increases pressure on the Bab el-Mandeb and the Cape of Good Hope. A shift in Arctic navigability transforms the strategic relevance of the Bering Strait. Global supply chains and maritime chokepoints behave as complex adaptive systems: independent agents—shipping companies, littoral states, insurers, navies—interact based on local incentives, producing nonlinear and sometimes unpredictable macro-behaviors. A localized shock triggers a homeostatic reaction: insurers reprice risk, carriers reroute, navies adjust posture, and the system searches for a new equilibrium. Interdependence is therefore not merely connective; it is adaptive, self-organizing, and nonlinear.

4. Multi-Domain Hybridity

Modern chokepoints are hybrid spaces where digital, economic, ecological, and civilizational systems intersect. Undersea cables, satellite networks, and AI-driven logistics overlay the physical geometry of straits. Climate change alters navigability. Migration reshapes the social environment. Cyber-nautical disruption introduces new forms of risk. Hybridity exposes a deep ontological clash: capital and data seek zero friction and instantaneity, yet they must manifest in physical reality—400-meter ships navigating 700-meter channels against currents and weather. Straits are generators of friction in globalization. Digital twins allow predictive stress-testing, while true redundancy increasingly requires backing up space-based systems with hardened terrestrial alternatives. As demand for low-latency compute grows, subsea and nearshore data-center concepts tie critical processing capacity to coastal jurisdictions, turning straits into locations where compute, connectivity, and sovereignty co-locate. Hybridity raises governance stakes: liability, environmental permitting, critical-infrastructure security, and the risk that fixed compute assets become dual-use targets.

5. Temporal Dynamism

Chokepoints evolve. Their significance changes with climate patterns, technological shifts, geopolitical realignments, and economic transformations. The Bering Strait is emerging; Hormuz may decline in relative importance as the energy transition accelerates; the Taiwan Strait is intensifying. Temporal dynamism reframes resilience as an evolving project rather than a static condition. Circular-economy strategies represent civilization's attempt to engineer its way out of geographic determinism, while nature-based coastal adaptation demonstrates that geophysical resilience is increasingly ecological and temporal rather than purely concrete.

6. Civilizational Encoding

Chokepoints carry symbolic meaning. They mark boundaries between cultural spheres, historical narratives, and political identities. The Bosporus embodies the legacy of empire. Gibraltar symbolizes sovereignty and continuity. The Taiwan Strait represents competing visions of political order. These meanings shape strategic behavior as much as material factors. A particularly acute human dimension lies in the maritime labor force: if crews refuse to sail into high-risk zones, the physical openness of a strait becomes irrelevant. Migration flows across Bab el-Mandeb, Gibraltar, and the Aegean further reshape littoral politics. To counter high-risk "shadow fleet" operations, states increasingly rely on Port State Control memorandums that reassert sovereign boundaries over the illusion of frictionless passage. Civilizational encoding embeds identity into geography.

Integrative Propositions of the Unified Theory

The unified theory of chokepoints can be distilled into three integrative propositions that together express the structural logic of the global system. The first proposition is that chokepoints are the structural nodes where global order becomes visible. They reveal the architecture of flows, dependencies, and power relations that normally remain abstract or distributed. In these narrow spaces, the scale-free nature of global networks becomes unmistakable: a small number of corridors carry a disproportionate share of global movement, and the unequal distribution of risk between host geographies and financial gatekeepers becomes impossible to ignore. Chokepoints expose the skeleton of globalization. They show where the world is most tightly bound together—and where it is most likely to break.

The second proposition is that chokepoints are the primary sites where systemic risk concentrates. Because they compress movement, they transform local disruptions into global consequences. A single blockage, accident, or act of coercion can propagate across continents, triggering nonlinear cascades in markets, supply chains, and strategic behavior. These cascades follow the logic of complex adaptive systems: insurers reprice risk, carriers reroute, navies reposition, and the system reorganizes itself around the disruption. The "reprice and reroute" dynamic is not a metaphor but a structural response pattern. Chokepoints are therefore not merely vulnerable; they are amplifiers. They convert friction into systemic shock.

The third proposition is that chokepoints are the leverage points through which future order will be shaped. Control, stabilization, and transformation of these nodes will determine the trajectory of global governance in the decades ahead. As societies confront the tension between digital instantaneity and physical constraint, chokepoints become the arenas where this tension is negotiated. They are where sovereignty is asserted, where environmental and technological standards are enforced, where civilizational narratives collide, and where the limits of global interdependence are tested. The ability to manage, secure, and redesign these narrow spaces will define which states and institutions can shape the emerging order—and which will be shaped by it.

Together, these propositions form the integrative core of the unified theory. They show that chokepoints are not peripheral features of the international system but central mechanisms through which power, risk, and order are produced. They are the places where the abstract becomes concrete, where the global becomes local, and where the future of the world system is most acutely negotiated. In these narrow spaces, the architecture of globalization reveals its true shape: a world defined not by openness but by compression, not by frictionless movement but by the management of constraint, not by uniformity but by asymmetry, hybridity, and temporal evolution.

This chapter completes the theoretical architecture of the book. What remains is the conclusion—the articulation of what it means to live in a world defined by narrow places, and how the chokepoint century will shape the evolution of global order.

24

Chapter 24

Assessing the Framework: Limits and Comparative Lessons

The Straits Systems Architecture (SSA) model was developed to provide a structured yet flexible framework for analyzing narrow maritime places as dynamic, multi-layered systems. It organizes the forces shaping chokepoint significance into six interdependent subsystems and treats each strait as a configuration of these subsystems rather than a static geographic feature. The model has proven useful in revealing patterns that cut across individual cases and in supporting the development of the Unified Theory of Chokepoints. Like any analytical framework, however, it has inherent limitations that must be acknowledged if the theory is to retain credibility and utility.

SSA Boundary Conditions

A unified framework also requires explicit boundary conditions. Not all constraints are chokepoints, and not all chokepoint language travels cleanly across domains. The SSA model is most useful when it is used to diagnose situations in which three conditions converge: (1) *compression* forces flows into a narrow set of predictable trajectories; (2) *dependency* is high enough that disruption materially changes system outcomes; and (3) *substitutability* is low enough that actors cannot cheaply or quickly reroute, replace, or postpone the flow.

Note on terminology: in this book, "optionality" and "substitutability" refer to the same underlying condition—whether credible alternatives exist at scale within the relevant time horizon and at acceptable cost.

Minimum compression threshold. A corridor stops behaving like a chokepoint when it no longer forces concentration into observable, governable trajectories. This does not require extreme narrowness on a map; it requires *operational* narrowing—where traffic separation schemes, draft limits, turns, shallow water, coastal surveillance, or engineered infrastructure constrain movement such that avoidance is difficult and interaction is compelled. Where flows can disperse across many equivalent routes without meaningful cost or exposure, the SSA model becomes less diagnostic because compression no longer produces leverage.

Substitutability ceiling. Chokepoint power weakens as substitutes become credible at scale. Substitutes can be geographic (alternate straits, Cape routing), infrastructural (pipelines, rail landbridges), technological (communication redundancy), or economic (inventory buffers, demand destruction). The relevant question is not whether an alternative exists in theory, but whether it can absorb the displaced flow at acceptable cost and within the time horizon that matters. When optionality is high and switching costs are low, disruption becomes an inconvenience rather than a systemic shock, and chokepoint leverage declines accordingly.

Time–elasticity limits. A chokepoint is most consequential when the system cannot tolerate delay. Some flows are highly time-elastic: they can be postponed, stored, substituted, or rerouted without catastrophic second-order effects. Others are time-inelastic: energy supply, just-in-time manufacturing, perishable and bio-industrial inputs, and crisis logistics can tip into non-linear failure when delayed beyond a threshold. In SSA terms, time-elasticity links economic flows to governance and technological subsystems, because delay often manifests first through insurance withdrawal, schedule desynchronization, and regulatory constraint rather than through physical closure.

Cross-domain caution (data, finance, air cargo). The SSA model can illuminate non-maritime chokepoints—such as undersea cable clusters, insurance withdrawal, or critical mineral processing—when those systems exhibit the same triad of compression, dependency, and low substitutability. But the model should not be used as a metaphor for every form of risk concentration. Where flows are purely virtual, instantly reroutable, or governed by many interchangeable nodes, chokepoint language becomes less precise and the SSA framework should be treated as a heuristic rather than a diagnostic tool.

- **Is movement forced into a narrow set of trajectories?** (Operational compression)
- **Would disruption change outcomes at system scale?** (De-

pendency)

- **Can the flow be replaced, rerouted, or delayed cheaply and quickly?** (Substitutability and time-elasticity)
- **What would failure look like first?** (Physical stoppage, financial withdrawal, temporal desynchronization, or governance breakdown)

Dual-use governance. Governance stabilizes chokepoints for those aligned with the regime that governs them—and weaponizes access against those outside it, often through regulatory conditioning, liability enforcement, and financial gatekeeping rather than overt closure.

Precision on transit passage. Environmental and insurance instruments do not nullify transit passage; they reprice and condition it—often by shifting the practical question from legal permission to commercial and operational viability.

Failure, Disruption, and Chokepoint Breakdown

A second clarity requirement is definitional. In this book, *disruption* refers to a measurable degradation of flow—higher risk pricing, slower transit, longer queues, partial rerouting, or episodic interdiction—that the system can still absorb without losing basic function. *Failure* refers to a threshold condition in which the corridor (or the system that depends on it) can no longer perform its core role at the relevant scale or time horizon. Failures can be physical, financial, temporal, or institutional, and in modern chokepoints they often occur in combination.

- **Functional failure (flow stops).** Physical movement through the corridor is halted or reduced below usable capacity by closure, obstruction, kinetic attack, sabotage, or environmental constraint.

- **Financial failure (coverage withdrawn).** The corridor remains physically open, but commercial transit becomes non-viable because insurance, credit, flag-state services, or essential contractual guarantees are withdrawn or priced beyond tolerance.
- **Temporal failure (system slows beyond tolerance).** Passage remains possible, but delay exceeds what dependent systems can absorb—triggering desynchronization of supply chains, inventory collapse, production stoppages, or cascading congestion across substitute routes.
- **Governance failure (rules unenforceable).** The corridor becomes strategically ambiguous because legal regimes, deconfliction mechanisms, or enforcement capacity break down—allowing coercion, predation, or miscalculation to dominate behavior even without formal closure.

These failure modes can cascade. Financial failure can precede functional failure (a strait becomes commercially "closed" before it is physically closed). Temporal failure can become functional failure when queues, congestion, or diversion exceed port and fleet capacity. Governance failure can trigger all three by lowering the threshold for interference and raising the probability of escalation. In SSA terms, functional failure is anchored in geophysical and strategic-military subsystems; financial failure is anchored in governance and technological infrastructures (insurance, data, and compliance); temporal failure is anchored in economic flows and optionality constraints; and governance failure emerges from the breakdown of rules, legitimacy, and enforcement across the entire architecture.

Labor withdrawal is a frequent trigger in contemporary disruptions: crew refusal, staffing shortages, or port labor constraints can convert heightened threat perception into real delay, and real delay into temporal failure—sometimes cascading into functional failure when congestion and schedule breakdown overwhelm available capacity.

The SSA model is primarily explanatory rather than predictive. **SSA is a structural heuristic, not a predictive model; it identifies failure modes rather than forecasting events.** It excels at illuminating why a particular strait matters, how its subsystems interact, and why disruption in one corridor can propagate through others. It is less effective at forecasting the precise timing, form, or political resolution of crises. This limitation stems from the model's structural orientation: it maps the architecture of vulnerability and leverage but cannot fully capture contingent political decisions, leadership calculations, or black-swan events that often determine whether potential disruption becomes actualized. A drought in Panama or military exercises around Taiwan may create the conditions for crisis, but whether those conditions produce systemic shock depends on choices made by actors whose behavior the model can contextualize but not determine.

A second limitation concerns the weighting and dynamism of the six subsystems. The SSA model treats the subsystems as interdependent, yet it does not prescribe fixed weights or causal hierarchies among them. In practice, the relative importance of geophysical constraints, military geometry, or technological infrastructures shifts over time and across cases. The model can identify which subsystems are most salient in a given strait at a given moment, but it offers limited guidance on how to anticipate or model shifts in their relative influence.

This reflects a deliberate choice to prioritize configurational analysis over causal modeling, but it also means the framework is better suited to comparative and diagnostic work than to formal prediction or simulation.

The model also faces a deeper challenge in capturing agency. While it can map how structural conditions shape the environment in which decisions are made, it struggles to account for the independent decision-making power of non-state actors whose choices can override the preferences of littoral states. In the contemporary chokepoint system, major shipping conglomerates and marine insurance syndicates frequently exercise effective veto power over physical geography. When these corporate actors determine that a corridor has become too risky and withdraw capacity or coverage, the strait can be functionally closed regardless of the political or military posture of the states that border it. The SSA model is not currently equipped to systematically weight this form of corporate sovereignty alongside traditional state agency, even though such decisions increasingly determine whether a chokepoint remains operationally open.

Data availability and comparability present a further constraint. While some dimensions—such as vessel transits, commodity flows, or military activity reports—are relatively well documented, others, particularly civilizational interfaces and the finer dynamics of hybrid technological risk, rely on more qualitative or fragmentary evidence. Insurance pricing, cyber incident reporting, and the operational details of digital infrastructure are often opaque or asymmetrically available. The model therefore operates with uneven empirical foundations, and any comparative claims must be tempered by awareness of what remains obscured or contested.

Even more fundamentally, the SSA model rests on an implicit assumption that large-scale maritime shipping will remain the primary mechanism of global material exchange for the foreseeable future. This assumption may not hold indefinitely. A true black-swan technological transformation—such as the widespread deployment of hyper-localized molecular manufacturing, advanced synthetic biology capable of producing materials and pharmaceuticals at point of need, or commercial fusion energy that drastically reduces the volume of raw materials requiring long-distance transport—could render much of today's oceanic trade obsolete. In such a scenario, the entire architecture of maritime chokepoints would lose its systemic centrality. The model is not designed to anticipate or analyze a post-maritime global metabolism, and this represents its most profound long-term boundary condition.

Finally, while the SSA model beautifully explains the existence of systemic interdependence among straits, it lacks the quantitative tools necessary to measure inter-strait elasticity with precision. It can identify that rerouting from Suez to the Cape of Good Hope increases pressure on alternative corridors and infrastructure, but it cannot currently forecast the specific thresholds at which bunkering capacity, port congestion, or insurance markets in those secondary routes will reach breaking point. This quantitative gap limits the model's utility for operational planning and invites future collaboration with economists, network scientists, and logistics modelers who can build elasticity and stress-testing layers on top of the qualitative framework developed here.

Despite these limitations, the SSA model enables a comparative analysis that reveals both common patterns and distinctive configurations across the world's principal maritime chokepoints. When the major cases are examined together, several structural contrasts become visible.

The straits with the highest structural compression and systemic leverage remain Hormuz and the Taiwan Strait. Both concentrate critical flows—energy in the case of Hormuz, advanced semiconductors in the case of Taiwan—within corridors where military geometry is intensely contested and where disruption would generate immediate, non-linear global effects. In both cases, dependency asymmetry is extreme.

A second group—Malacca, Bab el-Mandeb, and the Danish Straits—stands out for the intensity of their hybrid exposure. These corridors combine extreme commercial density with cable convergence, persistent gray-zone pressures, and the growing presence of shadow fleets. They illustrate how technological infrastructures and hybrid threats are reshaping the character of chokepoint risk even in corridors with relatively stable governance regimes.

The engineered corridors—Suez and Panama—share a distinctive profile of climate sensitivity and modality constraint. Both have experienced recent crises that forced large-scale rerouting and revealed the limited slack in global shipping capacity. Both are also sites where states are actively constructing alternatives, whether through land-bridges such as Mexico's Interoceanic Corridor of the Isthmus of Tehuantepec or through new port investments such as Chancay, in an effort to reduce exposure to single points of failure.

The Bosporus and Dardanelles occupy a distinctive position defined by high civilizational encoding and institutionalized but contested governance. They illustrate the Tragedy of the Chokepoint Commons in particularly stark form: the littoral state bears significant ecological and security burdens while the global benefits of passage accrue elsewhere.

The Bering Strait represents the clearest case of temporal emergence. Once marginal, it is becoming strategically and economically relevant as Arctic ice recedes. It combines geophysical transformation, great-power proximity, indigenous interfaces, and nascent hybrid infrastructure.

The Strait of Gibraltar occupies a category of its own, best understood as the archetype of historical continuity and modality constraints. As a natural gateway that has remained remarkably stable over centuries, it continues to function as a rigid geographic anchor between the Atlantic and the Mediterranean. Its contemporary dynamics are shaped by the interplay of longstanding sovereignty arrangements, the post-Brexit reconfiguration of European border regimes, and the persistent presence of Russian shadow fleet activity in its approaches. Gibraltar demonstrates how a chokepoint can retain deep historical and civilizational significance while still serving as a critical, non-substitutable interface whose disruption would impose severe modality constraints on both commercial and energy flows.

When these configurations are viewed together, the comparative logic of the SSA model and the Unified Theory of Chokepoints becomes clearest. Structural compression remains the foundational condition, but its consequences vary dramatically depending on the configuration of the other subsystems. Dependency asymmetry is most consequential where critical flows are concentrated and where few immediate substitutes exist. Systemic interdependence is most visible in the speed and scale with which disruption in one corridor forces adjustments across others. Multi-domain hybridity is no longer a secondary feature but a defining characteristic of the most systemically significant straits. Temporal dynamism reminds us that the map of chokepoints is not static. Civilizational encoding continues to shape political thresholds even in highly institutionalized or technologically dense environments.

The limitations of the SSA model do not invalidate these comparative insights; they clarify the conditions under which the model is most useful. The framework is strongest when deployed for structural diagnosis, comparative analysis, and the identification of patterns that cut across individual cases. It is weaker when asked to predict specific outcomes or to substitute for detailed political or historical reconstruction. Used with appropriate modesty, the model and the broader Unified Theory of Chokepoints provide a coherent language for understanding why narrow places have remained central to global order and why they are likely to remain so. The task for future work is not to overcome the model's limitations but to apply it with precision, to test its propositions against new cases and evolving conditions, and to remain attentive to the political and ethical questions that structural analysis alone cannot resolve.

25

Chapter 25

The Financial Strait: Insurance as the Hidden Chokepoint

The most consequential chokepoints in the twenty-first century are no longer only geographic. They are also financial. A corridor can remain physically open and yet become functionally closed if the contracts, credit, and liability protections that make transit commercially possible are withdrawn. In practice, the ability to insure a voyage is often the ability to sail it. Insurance therefore functions as a form of corridor governance—an enforcement layer that can open, condition, or deny access without any formal blockade. To understand chokepoint power in the modern era, insurance must be treated not as a background market but as a system actor.

CHAPTER 25

Insurance as Gatekeeper

Modern shipping is a liability business before it is a transport business. The core instrument is Protection and Indemnity (P&I) insurance, which covers third-party liabilities—pollution, collision, wreck removal, injury, and cargo claims—that can easily exceed the value of the ship itself. War-risk insurance sits alongside P&I as the mechanism that prices kinetic and hybrid threat, from missiles and mines to seizure and sabotage. Reinsurance then functions as the backstop that allows insurers to write catastrophic risk at scale. This layered structure means that "access" to a corridor is often a bundled product: physical navigability plus financeability.

Operational agility exists within constraints imposed by insurance, labor, and sovereign authority. Routing decisions, speed adjustments, and fleet redeployment can absorb some shocks, but they are bounded by financeability, crew willingness, port capacity, and enforceable rules of access.

This creates a distinctive chokepoint failure mode: *financial failure*. A strait can remain physically open while becoming commercially non-viable because P&I, war-risk coverage, or essential credit and contractual guarantees are withdrawn—or repriced beyond tolerance. In such moments, insurers and banks do what navies cannot: they compel immediate compliance by making transit unfinanceable. The result is a blockade-like effect generated through market governance rather than kinetic force, often preceding any physical interdiction.

The rise of InsurTech and AI-assisted underwriting compresses this dynamic in time. When risk models ingest satellite data, AIS anomalies, intelligence reporting, and claims history to update premiums in near real time, the chokepoint moves upstream: "closure" can occur as a pricing decision before any vessel is attacked or any state declares a blockade. In this sense, algorithms become corridor governors. They shorten escalation ladders by converting threat perception into immediate cost, and they create a new arena of power defined by data access, model credibility, and the ability to influence risk narratives.

As straits become substrates for fixed digital infrastructure, insurance exposure also changes in kind. Subsea, nearshore, and floating data-center deployments concentrate high-value compute and connectivity in coastal jurisdictions where outages, contamination events, or security incidents can generate business-interruption claims, environmental liability, and politically sensitive "critical infrastructure" losses. For underwriters, this is an aggregation problem: a single anchoring incident, sabotage event, or exclusion zone near a cable landing can create correlated losses across shipping, communications, and compute. The practical implication is that the liability architecture of chokepoints must now account not only for hull and cargo risk, but for contingent digital and infrastructure liabilities that make corridor financeability more fragile under stress.

Shadow Insurance and Parallel Logistics

Sanctions regimes have made insurance a central battlefield. As Western P&I clubs and reputable flag-state and classification services withdrew from sanctioned trade, an opaque, non-Western reinsurance and flagging ecosystem expanded to sustain "shadow" shipping. This does not eliminate risk; it displaces and amplifies it. Underinsured vessels operating with unclear ownership and weak compliance are more likely to generate catastrophic pollution and collision losses, and less likely to provide credible compensation when they do. The result is a bifurcated maritime economy: parallel "white" and "dark" logistical systems sharing the same physical corridors but operating under incompatible liability, transparency, and enforcement regimes.

This bifurcation sharpens the Tragedy of the Chokepoint Commons. Global markets benefit from continued flows, but littoral states inherit the downside: spill response, fisheries damage, legal uncertainty, and reputational risk when accidents occur in confined waters. In the Danish Straits, shadow-fleet passages elevate ecological risk in shallow channels; in the Red Sea approaches, cable cuts and hybrid attacks demonstrate how fragile corridors can become when enforcement is uneven. Shadow insurance therefore functions as a governance substitute that keeps commerce moving while weakening the liability architecture that normally internalizes risk.

Liability is therefore not an afterthought but part of corridor governance: without credible compensation and enforceable responsibility for pollution and wreck removal, insurance bifurcation turns chokepoints into zones of ecological moral hazard, shifting catastrophic cost onto littoral states while preserving global flow.

Sovereign Backstops and the Politics of Risk

When private insurers retreat, the only entity with a balance sheet large enough to keep corridors open is the state. Sovereign insurance backstops—explicit guarantees, war-risk reinsurance, indemnities for critical voyages, or state-supported pools—convert resilience from a market function into a public one. They can stabilize trade in acute crises, but they also create political choices: who is eligible for coverage, under what rules, and at what strategic cost. In effect, states can use backstops to preserve access for allies, deny access to adversaries, or condition passage on compliance with environmental and security standards. Finance becomes statecraft.

In SSA terms, insurance power sits at the intersection of governance regimes, technological infrastructures, and economic flows. It converts information into enforceable constraint, translating threat perception into either permission to sail or functional closure. It therefore clarifies a broader lesson of the chokepoint century: coercion is no longer only the product of missiles and mines. It is also the product of liability regimes, underwriting decisions, and the institutional capacity to absorb catastrophic risk. Any complete account of chokepoint resilience must treat financial failure as a primary failure mode, not a secondary effect.

Understanding insurance as a hidden chokepoint closes the loop between geography and governance. It shows why narrow places are simultaneously physical corridors and institutional constructs: passage is shaped as much by financeability as by navigability. The conclusion returns to the broader implications of this condition—ethical, political, and philosophical—and to what it means to live in a world whose order depends on narrow places.

26

Conclusion

The narrow places of the world do not merely punctuate global order. They constitute it. Across the preceding chapters, the argument has moved from the observation that certain maritime corridors exercise disproportionate influence over the circulation of energy, goods, data, and military power, to a systematic account of why this is so, and finally to a philosophical claim about what their persistence reveals about the nature of complex systems under conditions of compression. The Unified Theory of Chokepoints does not predict the next crisis. It explains why crises continue to emerge from the same structural features, why those features resist both technological dissolution and political erasure, and why the distribution of their costs and benefits remains so starkly unequal.

What it means to live in a world defined by narrow places is, first of all, to live with the consequences of structural compression. The same optimization that produces extraordinary efficiency in global supply chains also produces extraordinary fragility at their points of concentration. This is not a design flaw that can be engineered away. It is the mathematical signature of scale-free networks and the operational signature of Complex Adaptive Systems: when independent agents interact according to local rules under conditions of extreme concentration, non-linear and often unpredictable macro-behavior is the expected outcome rather than the exception. The chokepoint century is therefore not a temporary phase of heightened geopolitical tension. It is the condition that emerges when a hyper-connected world discovers that its connectivity has been purchased at the price of concentrated vulnerability.

One final horizon point follows from the book's argument. The Unified Theory of Chokepoints is a theory of the present regime of global material exchange—an era still governed by ships, corridors, and constrained geography. A true black-swan shift toward hyper-localized manufacturing or synthetic production could one day reduce oceanic throughput and weaken the systemic centrality of straits, but until such a transition occurs, narrow places will remain the decisive points at which global order is stressed and governed.

To live in this condition is also to confront an asymmetry that is at once material and ethical. The physical corridors through which global flows must pass are overwhelmingly located in the Global South. The ecological damage, the sovereign debt incurred in crisis response, and the political pressures that accompany militarization or great-power competition fall disproportionately on the states and communities that host these corridors. Meanwhile, the financial and regulatory architectures that determine who may transit, at what cost, and under what insurance regime remain heavily concentrated in the Global North. The Tragedy of the Chokepoint Commons is not a metaphor. It is a description of how risk and reward are allocated in a system that treats frictionless passage as a global public good while treating the costs of its disruption as a local private burden.

Policy Implications: Governing in the Chokepoint Century

The Unified Theory of Chokepoints and the Straits Systems Architecture model carry direct implications for how different actors should think and act. These implications are organized by the type of actor most directly concerned, while recognizing that effective governance requires coordination across categories.

For Littoral States Hosting Major Chokepoints

States that physically host critical straits face distinctive pressures. They bear disproportionate ecological, security, and infrastructural costs while often lacking the capacity to manage them alone. For these states, the central imperative is to convert geographic exposure into negotiated leverage. This requires treating environmental and infrastructural resilience as core elements of national security, developing the institutional capacity to engage corporate actors on equal terms, and pursuing minilateral arrangements with neighboring countries and key user states to share the burdens of security, environmental monitoring, and crisis response.

For Major Trading and Maritime Powers

For states whose economies and security depend heavily on multiple chokepoints, the central challenge is to reduce excessive concentration of risk while maintaining access. The most effective approach combines three elements: treating the development of credible alternative pathways as a strategic priority, investing in real-time situational awareness and predictive modeling capabilities that allow rapid adaptation, and developing new forms of public-private coordination that can align commercial risk management with broader strategic objectives.

For Middle Powers and Global South States

For states that are neither primary hosts nor dominant maritime powers, the chokepoint century presents both acute risks and potential opportunities. These states are often most exposed to secondary effects of disruption while possessing limited direct influence. The most promising strategy lies in collective action at the regional level through minilateral mechanisms focused on specific corridors or shared vulnerabilities, alongside support for strengthening the legal and normative foundations of the chokepoint commons.

For the Private Sector

Corporate actors—particularly major shipping companies and marine insurers—now exercise forms of authority over chokepoints that were once the exclusive domain of states. This power carries corresponding responsibilities. The most constructive role for the private sector is to move from reactive risk avoidance to proactive co-governance. This includes greater transparency in insurance and routing decisions, investment in shared situational awareness platforms, and participation in minilateral arrangements that bring together states, insurers, and operators around specific corridors.

Three priorities apply across all categories. First, the international community must develop more robust mechanisms for managing the Tragedy of the Chokepoint Commons, including new financial instruments that can absorb catastrophic shocks without forcing host states into unsustainable debt. Second, governance arrangements must be designed for hybrid environments in which physical security, cyber defense, environmental protection, and insurance stability are integrated into coherent frameworks. Third, all actors should treat the temporal character of chokepoint significance as a core planning assumption rather than relying on static strategies based on the current map of corridors.

Environmental Justice and Liability in the Chokepoint Commons

The Tragedy of the Chokepoint Commons raises a question that structural analysis alone cannot avoid: *who ought to pay* when global passage generates localized catastrophe? A defensible answer has two components. First is the classic *polluter-pays* principle: operators and cargo owners whose activity creates risk should bear the costs of prevention, response, and restoration. Second is a *beneficiary-pays* principle appropriate to a global commons: major user states and firms that derive systemic benefit from frictionless passage should contribute to the governance capacity and environmental protection of the corridors they depend on. Without this second component, liability regimes will remain formally correct but practically hollow in littoral states with limited response capacity and high exposure.

Normative responsibility must be operationalized through enforceable mechanisms. The first is strengthened *strict liability* and *mandatory financial security* for high-consequence corridors: credible P&I coverage, transparent ownership, verified classification status, and pre-committed spill response arrangements should be treated as conditions of access in the approaches to the most ecologically sensitive straits. Where historic free-passage regimes constrain direct tolling, enforcement can still occur through port-state control in destination and transshipment ports, through targeted inspections for unseaworthiness and insurance fraud, and through shared incident-response standards that make liability legible and collectible. The goal is not to suspend passage in normal times, but to prevent "liability-free navigation" in corridors where a single spill can become a national emergency.

The second tool is *international compensation*. If chokepoints are treated as shared infrastructure, then catastrophic environmental loss should not be financed solely through the sovereign debt of the host state. A dedicated "chokepoint environmental compensation facility"—funded by a mix of user-state contributions, shipping levies collected outside the strait (e.g., at major destination ports), and reinsurance-linked pools—would allow rapid payout for spill response, fisheries restoration, and coastal livelihood recovery. Complementing this, littoral states can treat chokepoint ecology as a component of *environmental sovereign risk*: contingent credit lines, parametric catastrophe instruments, and insurance-backed resilience bonds can provide immediate liquidity after a spill or major incident without forcing fiscal collapse. The ethical claim becomes a financial one: if the world demands predictable passage, it must help finance the liabilities that passage creates.

These obligations are most urgent under conditions of shadow shipping and insurance bifurcation. When vessels operate outside credible liability regimes, the commons becomes an extraction system: flows continue, but accountability vanishes. The response must therefore integrate environmental enforcement into the same toolchain that already governs access—insurance verification, port-state control, sanctions compliance, and real-time risk intelligence. Environmental justice in chokepoints is not a separate agenda from resilience; it is one of its preconditions. A system that cannot assign responsibility for its externalities cannot remain stable under stress.

The Weight of Constraint

The tension between openness and constraint that defines narrow places has long preoccupied thinkers concerned with the spatial foundations of power. Carl Schmitt observed that the sea has historically represented a realm of freedom and movement in contrast to the ordered, bounded character of land. Yet even the sea, he noted, eventually encounters points at which movement is forced into channels and where order must be imposed. Chokepoints are the sites at which this encounter becomes most visible—where the supposed freedom of maritime space is revealed as dependent on the control of particular passages. In this sense, they are not exceptions to Schmitt's *nomos* of the earth but among its clearest expressions: moments in which the spatial order of the world is both contested and made legible.

Fernand Braudel's account of the Mediterranean offers a complementary perspective. For Braudel, the sea was not merely a surface across which trade moved, but a structured environment whose rhythms—geological, climatic, and economic—shaped civilizations over the *longue durée*. The narrow places of the Mediterranean were not incidental features but constitutive elements of a systemic whole. The same logic applies to the global maritime system today: chokepoints are not simply bottlenecks in an otherwise fluid network. They are the points at which the slow structures of geography, climate, and accumulated infrastructure most visibly determine the possibilities of movement and exchange.

Manuel Castells' distinction between the "space of flows" and the "space of places" illuminates a more contemporary dimension of the same problem. The digital and logistical architectures of globalization aspire to create a space of flows in which distance and friction are minimized. Yet this space of flows continually collides with the space of places—specific locations where physical infrastructure, sovereignty, and ecological limits impose constraint. Chokepoints are the most dramatic sites of this collision. They are where the aspiration to seamless global circulation encounters the obdurate reality of narrow channels, contested sovereignty, and concentrated risk. The friction that results is not a failure of the network society but one of its structural conditions.

These perspectives, though developed in different historical and intellectual contexts, converge on a common recognition: that complex systems of movement and exchange do not dissolve spatial constraint but rather concentrate it. Narrow places are the points at which this concentration becomes most consequential—both materially and symbolically. They reveal that the order of the world is not produced through the elimination of limits but through their organization at particular sites.

The Deeper Stakes

The chokepoint century does not require the invention of entirely new forms of international cooperation. It requires the adaptation of existing capacities—state, corporate, and minilateral—to a structural reality that has become impossible to ignore. Yet this adaptation carries a deeper risk. If the Global North, faced with the accumulating liabilities of the chokepoint commons, chooses to abandon it through accelerated friend-shoring, sovereign closed-loop industrial systems, and technological decoupling, the oceans risk a form of balkanization in which heavily armed regional blocs secure their own corridors while leaving the remainder of the global commons under-governed and increasingly dangerous. In such a world, the poorest and most trade-dependent nations would find themselves not only bearing the localized costs of disruption but also excluded from the networks of exchange that once offered a path out of structural vulnerability.

The ultimate test of the chokepoint century is therefore not whether states and corporations can protect their own access. It is whether they can sustain the conditions under which narrow places continue to function as shared infrastructure rather than as instruments of exclusion or coercion. This requires accepting that leverage and vulnerability are two faces of the same structural property, that the distribution of their costs is a political choice rather than a geographic inevitability, and that the friction between digital aspiration and physical reality is not a problem to be solved but a condition to be inhabited with greater clarity and responsibility.

The task of statecraft in the chokepoint century is therefore not to master the narrow places of the world, which is impossible, but to understand them well enough to navigate the constraints they impose and to accept responsibility for the consequences they generate. That understanding begins with the recognition that the most powerful forces in the international system are often the least visible until they are not, and that the future will be shaped less by the vastness of oceans than by the narrow places where systems converge. In this sense, the chokepoint century is not only a description of a strategic environment. It is an invitation to think more rigorously about what kind of order we are willing to sustain—and at whose expense.

Epilogue

The world has always been shaped by its narrow places. Empires rose and fell at straits, religions crossed them, economies depended on them, and wars were decided by who could hold or bypass them. Yet the twenty-first century has made this ancient truth newly urgent. The global system is more interconnected, more fragile, and more dependent on compressed corridors than at any point in human history. The straits that once linked regions now bind the world.

This book began with a simple premise: that to understand the global order, one must understand the architecture of its chokepoints. But the deeper I went into the Straits Systems Architecture, the clearer it became that straits are not merely features of geography. They are mirrors. They reflect the systems we have built—our ambitions, our vulnerabilities, our interdependence, and our limits.

Hormuz reveals the world's dependence on energy and the fragility of its flows. Malacca reveals the density of trade and the cost of delay. Bab el-Mandeb reveals how local instability can scale into global disruption. Suez reveals the illusion of engineered certainty. Taiwan reveals the convergence of military, technological, and civilizational stakes. The Arctic reveals how climate change is rewriting the map faster than institutions can adapt.

Each strait is a system. Together, they form the skeleton of the global order.

The future will not be defined by the width of these corridors but by the choices we make around them. Will we treat straits as vulnerabilities to be exploited, or as shared infrastructures to be protected? Will we allow competition to harden into confrontation, or will we build governance architectures that match the complexity of the systems they must steward? Will we invest in resilience only after crisis forces our hand, or will we learn to anticipate the pressures that accumulate in narrow places?

The answers to these questions will shape the century.

The SSA model is not a prediction. It is a lens—a way of seeing the world as a set of interacting systems rather than isolated events. It is an invitation to think structurally, to recognize that the stability of the global system depends on understanding the architecture of its narrowest points. It is also a reminder that fragility and possibility often coexist. The same corridors that expose the world's vulnerabilities also enable its connections, its exchanges, and its shared futures.

In the end, straits teach us something profound: that the world is not held together by its vastness, but by its constraints. That power is not only exercised in open spaces, but in the places where movement tightens and choices narrow. And that the future will be shaped not by how wide the world is, but by how wisely we navigate its narrowest passages.

The world's straits will continue to define the flow of energy, trade, data, and ideas. They will continue to concentrate risk and opportunity. They will continue to reveal the architecture of global power. But they will also continue to remind us that systems—like straits—are shaped by human decisions.

The narrow places of the world are not just where the global system is most exposed. They are where it is most legible. And where its future will be decided.

Appendix

Applying the Unified Theory of Chokepoints (UTC)

One-Page SSA Recap

Use SSA to read any strait as a system composed of six interacting subsystems:

- **Geophysical determinants** — What physical constraints (depth, width, currents, ice, weather) compress movement and shape failure modes?
- **Strategic-military geometry** — Who can observe, deny, interdict, or escort, and what are the escalation ladders?
- **Economic and bio-industrial flows** — What critical flows (energy, containers, food inputs, industrial materials) depend on the corridor, and how time-inelastic are they?
- **Governance regimes** — What legal status, enforcement capacity, and institutional arrangements condition passage?
- **Demographic and civilizational interfaces** — What identity, migration, and social pressures shape legitimacy and escalation thresholds?
- **Technological infrastructures** — What cables, satellites, port OT, autonomy, and cyber dependencies create hybrid

vulnerabilities?

Quick test: chokepoint dynamics emerge when **compression**, **dependency**, and **low substitutability** converge. Diagnose how breakdown would appear first using the failure taxonomy: **functional** (flow stops), **financial** (coverage/credit withdrawn), **temporal** (delay beyond tolerance), or **governance** (rules unenforceable). Chokepoints often fail through cascades (financial or governance failure preceding physical stoppage).

How to Use This Book

- **Government and national security leaders:** Start with Chapters 1–3 (definitions + SSA), then Chapter 16 (global system), and Chapters 17–20 (future shock, risk geography, statecraft, resilience). Use Chapter 24 for boundary conditions and failure definitions.
- **Military and coast guard planners:** Focus on Chapters 2–3 (mechanisms + SSA), the relevant case chapter(s) for your theater, Chapter 18 (statecraft), and Chapter 21 (operationalizing resilience). Treat Chapter 25 (insurance) as an operational constraint, not a background factor.
- **Port authorities and maritime administrations:** Use Chapters 3–5 (SSA + tech), your strait case chapter, then Chapters 20–21 (resilience design and mechanisms). Chapter 24 provides a diagnostic checklist for prioritizing investments.
- **Insurers, reinsurers, and regulators:** Start with Chapter 24 (failure modes) and Chapter 25 (The Financial Strait). Use Chapter 16 for systemic linkage effects and Chapters 20–21 for resilience mechanisms such as sovereign backstops, port-state control, and hybrid infrastructure protection.

- **Corporate supply-chain and energy risk teams:** Start with Chapters 2–3 (mechanisms + SSA), then Chapter 16 (system linkages) and the case chapters most relevant to your routes. Use Chapters 20–21 for operational mechanisms (redundancy, buffering, distributed capacity).

Rapid Chokepoint Triage Checklist (Decision-Tree in Words)

1. **Define the corridor.** What exact passage, approach, or infrastructure cluster is at issue (strait, canal, cable landing zone, port complex)?
2. **Define the critical flow.** What is moving (oil, LNG, containers, grain, semiconductors, data), and what is its time sensitivity?
3. **Test the chokepoint triad.** Are compression, dependency, and low substitutability all present? If one leg is missing, treat it as a constraint, not a chokepoint.
4. **Identify the primary failure mode.** Would breakdown appear first as functional, financial, temporal, or governance failure?
5. **Check leading indicators.** Look for signals such as insurance premium spikes or withdrawal, AIS anomalies/GNSS interference, port congestion, crew refusal, regulatory tightening, or escalatory patrol patterns.
6. **Select the first-line response.** Choose the mechanism that matches the failure mode: buffering for temporal shocks, sovereign backstops for financial shocks, escorts/deconflic tion for governance breakdown, and hybrid infrastructure protection for cyber/cable threats.
7. **Assess spillover.** Where will rerouting pressure relocate

(Cape arc, alternate straits, inland corridors), and what second-order risks are created?

Failure-Mode Playbook (Where to Look Next)

- **If failure is functional (flow stops):** prioritize physical clearance, escort, deconfliction, and emergency rerouting. Re-read the relevant case chapter and Chapters 2 and 21; use Chapter 16 to anticipate spillover corridors.
- **If failure is financial (coverage withdrawn):** treat insurance as the binding constraint. Use Chapter 25 first, then Chapters 21 and 24 for sovereign backstops and compliance enforcement tools (PSC, sanctions, liability).
- **If failure is temporal (delay beyond tolerance):** focus on buffering, distributed capacity, and schedule redesign. Use Chapters 20–21, and use the case chapters to estimate where congestion will relocate.
- **If failure is governance (rules unenforceable):** prioritize crisis communication channels, rules clarification, coalition coordination, and calibrated enforcement. Use Chapters 18 and 21, plus the relevant case chapter for legal regime specifics.

Decision Thresholds (Watch → Warn → Act → Escalate)

Use the four-step ladder below to translate diagnosis into action. The thresholds are deliberately simple: they are designed to align technical signals (AIS, GNSS, VTS, cable alerts) with market signals (insurance, charter rates, credit) and political signals (rules, patrol posture, declarations).

- **WATCH** — Early anomalies present; disruption is possible but not yet self-sustaining. *Typical triggers:* localized GNSS interference; single cable incident; isolated harassment; modest premium widening; early weather/drought indicators.
- **WARN** — Risk is repriced and behavior begins to change; delays and compliance friction start to accumulate. *Typical triggers:* war-risk premium spikes; P&I documentation tightening; early crew refusal signals; VTS congestion; advisory rerouting by major carriers.
- **ACT** — Failure becomes plausible within days/weeks unless mitigated; coordinated measures are required. *Typical triggers:* insurer withdrawal or exclusions; sustained GNSS degradation; repeated attacks/boardings; quarantine-style inspection regime; lock rationing; port OT/cyber disruption; multi-day queueing.
- **ESCALATE** — Failure is occurring or imminent; the system is now in cascade management. *Typical triggers:* physical closure/obstruction; credible mining/kinetic interdiction; widespread crew refusal; sovereign backstop activation; coalition escort operations; emergency stock releases; large-scale diversion producing zero-slack conditions.

At each threshold, explicitly name the primary failure mode (functional, financial, temporal, or governance). That label should determine the first-line response: sovereign backstops and compliance enforcement for financial failure; buffering and distributed capacity for temporal failure; escorts, deconfliction, and rules clarification for governance failure; and physical clearance, protection, and emergency rerouting for functional failure.

Cross-Stakeholder Responsibility and Lead-Agency Handoffs

- **Trigger detection (WATCH/WARN):** typically commercial and technical actors first—insurers, ship operators, VTS authorities, cable owners, and maritime intelligence providers—because they see repricing, anomalies, and congestion earliest.
- **Lead authority (ACT):** the coastal state (maritime administration + coast guard/navy) leads in its waters; major user states support through intelligence, escort capability, and diplomatic coordination.
- **Coordination function:** minilateral groupings and standing mechanisms (regional patrols, VTS coordination, incident response protocols) provide the fastest coordination; ad hoc coalitions fill gaps when institutions are weak.
- **Liability and compensation:** operators and cargo interests bear primary liability through P&I and contractual regimes; when those regimes fail (shadow shipping, insurer withdrawal), the host state absorbs first-order costs unless compensation facilities or sovereign backstops exist.
- **De-escalation and rules clarity:** foreign ministries and defense ministries must coordinate early to prevent governance failure from turning into kinetic escalation; deconfliction channels should be treated as a core resilience asset.

Operational Ladders (What to Do Next, in Sequence)

A. Military / Coast Guard ladder (governance and functional failure focus)

1. **ISR surge and fused MDA.** Increase persistence (air/surface/space), fuse AIS/radar/RF/satellite feeds, and publish advisory products to reduce uncertainty.
2. **Deconfliction and communications.** Activate hotlines and operational deconfliction procedures; issue clear rules of interaction to lower miscalculation risk.
3. **Escort and routing discipline.** Implement convoy/escort options and prioritized lanes when threat is persistent; coordinate with industry on timing windows.
4. **Inspection and compliance (when lawful).** Use coast-guard style inspections for safety, insurance, and sanctions compliance where legal basis exists; treat quarantine-style coercion as a governance failure trigger.
5. **Denial and protective operations.** If interdiction becomes kinetic, shift to active defense (air defense, mine countermeasures, interdiction of launch platforms) while maintaining merchant protection.
6. **Coalition escalation and burden-sharing.** Formalize combined tasking, clarify who pays and who commands, and align sovereign backstops with escort posture to keep transit financeable.

B. Port / VTS / maritime administration ladder (temporal and financial failure focus)

1. **Stabilize traffic flow.** Tighten reporting compliance, sequencing, and speed/spacing rules; publish queueing and anchorage protocols.
2. **Surge safety capacity.** Increase pilotage windows, tug availability, salvage readiness, and emergency tow capability in the approaches.

3. **Harden cyber-nautical systems.** Raise OT security posture (access controls, incident response, segmented networks) and validate backup communications.
4. **Activate spill and casualty response.** Pre-position containment, coordinate mutual aid, and verify contractor availability for high-consequence incidents.
5. **Escalate port state control (PSC).** Target high-risk vessels (insurance gaps, ownership opacity, unseaworthiness indicators) before they enter constrained waters or as they exit into controlled anchorages.
6. **Contingency capacity and diversion planning.** Pre-negotiate overflow berth plans, alternate port calls, and cargo prioritization rules to reduce desynchronization cascades.

Role-Specific Playbooks (One-Page Each)

State (national security and economic ministries)

- Declare the failure mode early (functional/financial/temporal/governance) to align agencies and markets.
- Pre-authorize emergency authorities for escorts, inspections, and sanctions enforcement.
- Coordinate sovereign backstops (war-risk guarantees, indemnities) with operational posture so financeability matches physical protection.
- Activate commodity and industrial buffers (SPR, gas storage, food reserves) with clear triggers and communication.
- Engage littoral partners with burden-sharing offers (spill response, ISR, capacity funding) rather than demands for "free passage."

Military and coast guard commands

- Prioritize persistent MDA and early warning; publish shareable advisories to reduce uncertainty.
- Separate deconfliction tasks from coercive tasks; protect communications channels as a stability asset.
- Use white-hull tools (inspections, safety enforcement, escort discipline) to manage gray-zone coercion without uncontrolled escalation.
- Plan for mine countermeasures, air defense, and casualty response as core chokepoint missions.
- Coordinate coalition rules of engagement and cost-sharing before the corridor reaches ESCALATE.

Port authorities and maritime administrations

- Invest first in high-return safety capacity: VTS modernization, tug/salvage readiness, spill response, and OT cyber hardening.
- Pre-negotiate overflow anchorage and berth plans with neighboring ports to prevent queue cascades.
- Use PSC targeting to screen high-risk vessels (insurance gaps, ownership opacity, unseaworthiness) before they enter constrained waters.
- Operationalize real-time stress testing (digital twins where mature) to optimize sequencing and reroute decisions.
- Publish clear incident and compensation protocols to preserve legitimacy and reduce panic behavior.

Insurers, reinsurers, and regulators

- Distinguish clearly among *repricable* risk (premium and conditions), *excludable* risk (limited withdrawal), and *uninsurable* risk (systemic loss likelihood).
- Signal conditions for continued coverage (convoy windows, verified compliance, vetted ownership/class) to avoid abrupt market panic.
- Coordinate with states on sovereign backstop triggers to keep essential flows financeable.
- Share anonymized risk intelligence and claims patterns with maritime authorities to improve prevention.
- Treat shadow shipping as a liability regime problem: enforce financial security and transparency as conditions of access to coverage.

Corporate supply-chain and energy risk teams

- Map exposure to a manageable set of corridors (top 10 by value and time sensitivity), not "every chokepoint."
- Define maximum tolerable delay by product class and convert it into reroute and inventory triggers.
- Pre-contract alternative routings and ports; treat Cape routing as a capacity shock, not a solution.
- Maintain buffers where time-inelasticity is high (critical components, fuels, food inputs).
- Track leading indicators: insurance repricing, AIS/GNSS anomalies, queueing, and regulatory tightening; trigger action before physical closure.

What Not to Do (Common Failure Patterns)

- **Do not over-militarize a governance problem.** In many crises, the binding constraint is financial or legal; gray-hull escalation can worsen insurance withdrawal.
- **Do not ignore insurance and labor.** P&I withdrawal and crew refusal can close a corridor faster than physical interdiction.
- **Do not treat rerouting as a "solution."** Cape diversion is a capacity sink that creates second-order congestion and emissions shocks.
- **Do not externalize ecological risk.** Undercompensated spills create legitimacy crisis and retaliation incentives in littoral states.
- **Do not wait for physical closure.** Act at WARN/ACT thresholds; many chokepoints fail financially before they fail physically.

This appendix is designed as a navigation aid, not a substitute for the case studies. Its purpose is to let readers identify what kind of chokepoint problem they face, what "failure" would mean in that context, and which parts of the book contain the relevant tools, precedents, and concepts to respond with speed and coherence.

Glossary of Key Terms

Alternative PNT (A-PNT): Positioning, Navigation, and Timing systems that do not rely on satellite signals. Terrestrial systems such as eLoran provide hardened, low-frequency alternatives that are resistant to jamming and spoofing, offering redundancy when space-based systems are compromised.

Civilizational Encoding: The process by which narrow places acquire symbolic and historical meaning that shapes political behavior and strategic thresholds. Chokepoints are not only material corridors but sites where identity, memory, and narratives of sovereignty converge.

Complex Adaptive Systems (CAS): Systems composed of independent agents that interact according to local rules, producing emergent, non-linear, and often unpredictable macro-level behavior. Maritime chokepoints and global supply chains exhibit CAS characteristics, particularly in their capacity for rapid self-reorganization following localized shocks.

Corporate Statecraft: The exercise of effective authority over physical infrastructure and movement by non-state corporate actors—particularly major shipping lines and marine insurance syndicates—whose risk and coverage decisions can functionally open or close maritime corridors.

Dependency Asymmetry: The uneven distribution of reliance on chokepoints across states and economies. Some actors depend on particular straits for survival or critical functions, while others possess greater capacity to absorb or redirect flows, creating structural imbalances in vulnerability and leverage.

Digital Twins: High-fidelity, real-time virtual replicas of physical ports, straits, or logistics networks used for predictive stress-testing and operational optimization. Digital Twins allow operators to simulate disruptions before they occur.

Engineered Chokepoint: A narrow maritime corridor created or heavily modified by human intervention (such as the Panama Canal or Suez Canal), whose functionality depends on sustained infrastructure, water management, and technological systems rather than solely on natural geography.

Global South / Global North Disconnect: The structural

tension in the contemporary chokepoint system whereby the physical corridors and their ecological and sovereign risks are disproportionately located in the Global South, while the financial, insurance, and regulatory architectures that govern access remain concentrated in the Global North.

Minilateralism: Pragmatic, issue-specific cooperation among a limited number of capable actors, as opposed to universal multilateral institutions. In chokepoint governance, minilateral arrangements have emerged as the dominant form of coordination for security, environmental monitoring, and crisis response.

Multi-Domain Hybridity: The condition in which chokepoints are shaped by the interaction of physical, digital, economic, ecological, and civilizational systems. Modern chokepoint risk increasingly arises from the convergence of these domains rather than from any single dimension.

Nature-Based Solutions (NBS): The use of ecological processes—such as the restoration of wetlands, mangroves, and oyster reefs—to enhance coastal and chokepoint resilience. NBS can absorb wave energy, manage sedimentation, and provide cost-effective alternatives or complements to traditional "gray" infrastructure.

Ontological Clash (Digital Speed vs. Physical Drag): The fundamental tension between digital and algorithmic systems, which aspire to frictionless, instantaneous movement, and the geophysical realities of maritime space, where steel hulls, ocean currents, and narrow channels impose unavoidable constraint and friction.

Port State Control (PSC): The authority of port states to board, inspect, and detain foreign-flagged vessels for safety, labor, environmental, or security non-compliance. PSC serves as a key enforcement mechanism against shadow fleets and substandard shipping in chokepoint approaches.

Scale-Free Network: A network structure in which a small number of highly connected hubs account for a disproportionate share of connections. Global maritime trade exhibits scale-free characteristics, making chokepoints mathematically inevitable rather than accidental features of the system.

Shadow Fleet: Vessels engaged in illicit or gray-zone shipping, often older tankers operating without standard insurance, classification, or transparent ownership. Shadow fleets exploit jurisdictional ambiguities in chokepoints and have become a significant vector of hybrid risk.

Sovereign Insurance Backstop: State-provided or state-supported insurance mechanisms that underwrite war-risk or catastrophic coverage when private marine insurance markets withdraw from a corridor. Backstops function as financial buffers that prevent the functional closure of chokepoints during periods of extreme stress.

Straits Systems Architecture (SSA): An analytical framework that examines each chokepoint as a dynamic system composed of six interdependent subsystems: geophysical determinants, strategic-military geometry, economic and bio-industrial flows (including the human labor and operational capacity required to move ships and operate ports), governance regimes, demographic and civilizational interfaces, and technological infrastructures.

Structural Compression: The concentration of movement, risk, and leverage within narrow corridors. Compression is the foundational condition that transforms geography into strategic significance.

Systemic Interdependence: The condition in which disruption or change in one chokepoint generates effects across others through the interconnected architecture of global flows. Interdependence transforms local events into systemic variables.

Temporal Dynamism: The recognition that the significance of chokepoints is not fixed but evolves in response to climate change, technological development, geopolitical realignment, and economic transformation.

Tragedy of the Chokepoint Commons: The structural inequity whereby the benefits of frictionless global passage through narrow places are broadly distributed, while the ecological, security, and sovereign costs are disproportionately borne by the states and communities that host those places.

Unified Theory of Chokepoints (UTC): A theoretical framework built on six principles—structural compression, dependency asymmetry, systemic interdependence, multi-domain hybridity, temporal dynamism, and civilizational encoding—that explains the persistence and systemic role of narrow places in global order.

About the Author

Ali Ayoub, PhD, is a biotech inventor and strategist with over two decades of global experience. An American citizen born near Beirut, he brings an international perspective to the intersection of science and global security. Dr. Ayoub earned his PhD in Material Chemistry from the University of Reims and held fellowships at Cornell University and Japan's NARO.

A multidisciplinary strategist, he holds an MBA from the London School of Economics and has studied at Oxford and King's College London's Department of War Studies. He is the author of multiple books and research articles, and has developed proprietary frameworks and patents aimed at securing global systems.

As founder of Ayoub Sciences LLC, he advises the agribusiness sector on fortifying supply chains against geopolitical disruption using AI and sustainable innovation. He is guided by the belief that innovation must foster prosperity, advance human security, and build a sustainable future for all.

www.ingramcontent.com/pod-product-compliance
Lightning Source LLC
LaVergne TN
LVHW010651110826
845149LV00014B/3041

* 9 7 8 1 9 7 2 0 3 3 2 4 1 *